Selected Guidelines for Ethnobotanical Research:

A Field Manual

Selected Guidelines for
Ethnobotanical Research:
A Field Manual

Edited by Miguel N. Alexiades
with assistance from Jennie Wood Sheldon

NYBG

Scientific Publications Department
THE NEW YORK BOTANICAL GARDEN
Bronx, New York 10458-5126 U.S.A.
(718) 817-8721 • FAX (718) 817-8842
E-mail: scipubs@nybg.org

FLIP

Published by
**The New York Botanical Garden
Bronx, New York**

08 07 06 05 04 03 02 01 00 99 98   9 8 7 6 5 4 3 2

*Composition by Maple-Vail Composition Services
Cover photographs by Miguel N. Alexiades
Cover/interior design by Joy E. Runyon
Manufacturing by BookCrafters, Inc.*

The paper used in this publication meets the requirements of the American
National Standard for Information Sciences — Permanence of Paper for
Publications and Documents in Libraries and Archives (ANSI/NISO Z39.48–1992).

Printed in the United States of America using soy-based ink on recycled paper.

Metropolitan Life Foundation is a leadership funder of
The New York Botanical Garden's Scientific Publications Program.

Library of Congress Cataloging-in-Publication Data
Selected guidelines for ethnobotanical research : a field manual / Miguel
N. Alexiades, editor ; with assistance from Jennie Wood Sheldon.
    p. cm. — (Advances in economic botany ; v. 10)
    Includes bibliographical references and index.
    ISBN 0-89327-404-6
    1. Ethnobotany — Research — Handbooks, manuals, etc. I. Alexiades,
    Miguel N., 1962–   . II. Sheldon, Jennie Wood, 1959–   .
GN476.73.S453 1996
306.4'5'072 — dc20   96-18949

Contents

Foreword

Ethnobotany has developed greatly during the last two decades and has now become a recognised scientific discipline in its own right. The diversity of approaches that are mentioned in this volume point up the multidisciplinary nature of the science: In order to be successful, botanists should have an understanding of anthropology and anthropologists should have a knowledge of botany. (Indeed, many students and researchers in ethnobotany will find the bibliographic material given in the appendices of great use in broadening their base of knowledge.)

This most useful manual addresses three of the most important aspects of ethnobotanical research. First, although ethical issues have always been important, only recently has there been a concerted effort to respect the ownership rights of the knowledge of indigenous and local peoples. It is to be hoped that future researchers adhere to the guidelines provided here as well as to the norms of the Biodiversity Convention.

Second, much past research in ethnobotany, even that of some botanists, has been invalidated by the failure to collect adequate voucher specimens. This situation would be markedly improved if all future fieldworkers adhere to the guidelines discussed in this volume. A particularly useful feature of this manual is that it also includes instructions for the collection of the often-neglected fungi and bryophytes and the hard-to-collect palms.

Third, this manual has three useful chapters on the quantification of ethnobotanical data. Use of the methods outlined—and, indeed, of other new techniques—to quantify data will be an important part not only of encouraging the emergence of new ideas but also of demonstrating that ethnobotany is a science. Such data quantification and analysis, one of the more recent approaches to ethnobotany, supplant what formerly was just the compilation of lists of plants used or observations of behaviour and how plants are used.

The world still lacks any formal undergraduate- or graduate-level program in ethnobotany. It is to be hoped that this manual and other recently published guides for ethnobotany will stimulate a few centres of education to rise to the challenge. Here is a textbook that will greatly facilitate the teaching of the subject, that is also an essential part of the equipment of anyone embarking on ethnobotanical fieldwork.

Ghillean T. Prance, FRS
Director
Royal Botanic Gardens, Kew

Acknowledgments

We are grateful to Michael Balick, Douglas Daly, Christine Padoch, and Jan Stevenson for their encouragement in putting together and publishing a set of methodological guidelines. Special thanks to Joy E. Runyon for her invaluable assistance. We also thank Sandi Frank, Christine Padoch, and Charles Peters for their advice and support in editing. Two anonymous reviewers provided useful input that helped focus the approach, presentation, and contents of the manual. Two additional anonymous reviewers provided valuable input on a number of contributions. Robin Goodman assisted in the collection of bibliographic and reference material. We alone, however, are responsible for any omissions or errors. Additional reviewers and contributors are thanked at the end of several individual papers. The Institute of Economic Botany of The New York Botanical Garden and the National Cancer Institute provided financial support for the writing of several papers and the compilation of the manual.

Introduction

Miguel N. Alexiades

Institute of Economic Botany,
The New York Botanical Garden

Ethnobotany has been defined as "the study of the direct interrelations between humans and plants" (Ford, 1978a: 44). Because most ethnobotanical studies have emphasized the uses of plants by hunter–gatherer and agricultural societies, it is often assumed that ethnobotany is restricted to those communities. Actually, ethnobotany encompasses the study of all human societies, past and present, as well as all types of interrelations: ecological, evolutionary, and symbolic. Thus, ethnobotany recognizes the reciprocal and dynamic nature of the relationship between humans and plants (Ford, 1978a; Rosas, 1975, in Barrera, 1982). The term *economic botany* is sometimes used as a synonym for *ethnobotany,* though generally the former is used to encompass indirect as well as direct use of plants by humans (Wickens, 1990). Thus, ethnobotany is often considered a subfield of economic botany, though several authors have underscored the epistemological and philosophical differences between these two fields of study (e.g., Argueta et al., 1982; Barrera, 1982).

As Ford has indicated, "Ethnobotany lacks a unifying theory, but it does have a common discourse" (1978b: 29). Because plants play an important role in almost every realm of human activity, ethnobotany encompasses many fields, including botany, biochemistry, pharmacognosy, toxicology, medicine, nutrition, agriculture, ecology, evolution, comparative religion, sociology, anthropology, linguistics, cognitive studies, history, and archaeology. The multidisciplinary nature of ethnobotany allows for a wide array of approaches and applications, and it presents a challenge to researchers approaching the field from any one discipline. For example, botanists recording plant uses have often failed to document cultural data in a meaningful or consistent way, whereas "people-based" studies related to plant use have often neglected important biological aspects, such as the collection of plant specimens for positive taxonomic identification or the description of important ecological variables (Prance, 1991).

Within ethnobotany, Berlin (1992) recognizes two distinct approaches. The first, cognitive ethnobotany, deals with how humans view and classify plants; the second, economic ethnobotany focuses on how humans utilize plants (Berlin, 1992). Although the two approaches are clearly related, the former has been pursued primarily by linguists and anthropologists, whereas economic ethnobotany has been the focus of a much broader range of specialists, including botanists, archaeologists, anthropologists, geographers, pharmacologists and biochemists, physicians, foresters, and ecologists. Unlike cognitive studies, which often have employed a theoretical framework and involved the application of rigorous data collection and analysis techniques (see for example Atran, 1985; Berlin et al., 1974; Ellen & Reason, 1979; Hays, 1974; Hunn, 1982), economic studies have often remained purely descriptive and limited to the compilation of lists of plants. Although there are many notable exceptions (see, for example, Alcorn, 1984; Balée, 1994; Bodley & Benson, 1979; Browner, 1985; Conklin 1954; Cunningham & Mbnekum, 1993; Denevan & Padoch, 1987; Etkin, 1992; Harris & Hillman, 1989; Hernandez, 1970; Johns, 1990; Lewis & Elvin-Lewis, 1977; Matossian, 1989; Moerman, 1979; Padoch, 1990; Posey & Balée, 1989; Reichel-Dolmatoff, 1989; Salick, 1992; Schultes et al., 1977; Turner, 1988), all too often, studies in economic ethnobotany have failed to present the data within a broader framework that incorporates the evolutionary, pharmacological, ecological, cultural, historical, or social context of human–plant interactions (Barrera, 1982; Prance, 1991). Added to, or perhaps because of, this shortcoming, there has been a concomitant lack of methodological rigor in many studies in economic ethnobotany. These factors, along with the intrinsic resistance within "orthodox" sciences to accept disciplines that challenge traditional boundaries between academic disciplines, may have contributed to the perception among some skeptics of economic ethnobotany as a "pseudoscience" (Phillips & Gentry, 1993).

As part of the scientific inquiry, and in order to develop a sound methodology, ethnobotanists need to define the research problem, select a conceptual model, operationalize the chosen variables, and choose adequate field techniques (Pelto & Pelto, 1978). Because these decisions necessarily determine how and what data will be collected, they need to be made at the outset of the research. This manual seeks to assist ethnobotanists in one basic aspect of this methodological process: identifying and implementing field techniques suitable to a research question and field site. To this end, a broad range of applicable concepts and field techniques from anthropology, ecology, and botany are presented, and their use and relative merits are discussed.

The broad range of questions and approaches available in the study of plant–people interactions precludes a single comprehensive treatment. These guidelines are restricted to the subfield of economic ethnobotany. Martin (1995) presents a concise and useful overview of techniques related to cognitive ethnobotany. Researchers can also consult the extensive body of literature pertaining to cognitive anthropology and ethnoscience for detailed discussions of the concepts and techniques used to elucidate how local plant taxonomies are structured (e.g., Tyler, 1969; Weller & Romney, 1988; Werner & Schoepfle, 1987). Although this manual does not explicitly address the specific needs of ethnoecological and archaeoethnobotanical fieldwork, many of the concepts and techniques discussed will be applicable to these subfields of economic ethnobotany.

It is clear that inter- and multidisciplinary approaches can lead to more fruitful, thorough, and systematic approximations in the study of plant–people interactions (Elisabetsky, 1986; Etkin, 1986; Schultes & von Reis, 1995). Students thus need to gain a basic understanding of the paradigms, concepts, and techniques used by practitioners of the natural and social sciences, a process that unfortunately is not facilitated by the artificial barriers that frequently exist between different disciplines and academic programs. By drawing upon the experience of anthropologists, ethnobotanists, ecologists, foresters, and systematists, this manual

seeks to help willing students cross the conceptual bridges between disciplines in order to create broader perspectives and assist in the development of an ethnobotany for the future (Prance,
1995). The need for such guidelines is increasing as more researchers from a wide range of disciplines are drawn into ethnobotany and as the role of ethnobotany continues to grow in
health care (Akerle, 1988; Balick, 1990; Farnsworth et al., 1985;
Lozoya, 1976; Robineau & Soejarto, 1996), development (Alcorn, 1995; Bennett, 1992; Posey, 1982), and conservation (Phillips et al., 1994; Plotkin, 1995; Prance et al., 1987; Redford &
Padoch, 1992). In this sense, the present guidelines build upon
the growing, but still inadequate, body of literature on ethnobotanical field methods (see also Bellamy, 1993; Etkin, 1993; Given
& Harris, 1994; Jain, 1989; Lipp, 1989; Martin, 1995).

Although the principles and techniques discussed in this volume are applicable to most contexts of ethnobotanical research,
many of the examples are drawn from the tropics. This bias
stems from the geographical area of specialization of several of
the authors and should not detract from the value of the manual
to ethnobotanists working in other regions, since most of the
principles discussed will apply to most research contexts.

Ethnobotanical fieldwork presents not only theoretical and
methodological challenges but ethical challenges as well (Boom,
1990; Brush & Stabinsky, 1996; Elisabetsky, 1991; Posey, 1990;
Thrupp, 1989). Ethnobotanists collect both cultural and biological resources, and in many cases they transfer these resources between cultures, societies, or nation-states. The very nature of
much applied or economic ethnobotanical research implies that
ethnobotanists often intentionally or unintentionally serve as brokers of cultural and genetic resources. As Brush explains, this
role becomes problematic when "knowledge is freely shared in
one culture and then commoditized for private profit in another"
(1993: 653). Attempts to couch ethnobotany within the self-serv-
ing mythology of a value-free science or the "for-the-good-of-
humanity" discourse clearly are not tenable, particularly in light
of ethnobotany's historical role in the expansion of Western im-
perialism in the tropics (Brockway, 1979) and in the context of
the inequalities that have traditionally permeated models of exchange between temperate and tropical nations (Toledo, 1986).
These perspectives place a tremendous ethical burden on ethno-

botanists, who often work under serious time and financial constraints and at the interface of conflicting perceptions, interests, and needs. These issues, however, need to be continually addressed if the field is to develop in a rapidly changing world (Toledo, 1995).

For convenience, this manual has been divided into four parts. Part I deals with some of the logistical, methodological, and ethical aspects of ethnobotanical fieldwork, including the collection of cultural data. Chapter 1 discusses the protocols related to initiating and conducting ethnobotanical fieldwork, including obtaining permits, discussing and negotiating research objectives and priorities with communities, and selecting informants.[1] The question of ethics in ethnobotanical research, together with the issues of compensation and intellectual property rights, are systematically explored in Chapter 2. Chapter 3 introduces concepts and techniques central to the collection of cultural knowledge. This approach, based on the compilation of emic ethnobotanical knowledge, is complementary to the etic approaches discussed in Part III, which deal with ethnobotanical behavior.

Part II reviews the techniques used to collect plant specimens. Chapters 4 through 7 deal with the collection of voucher specimens, necessary to obtain adequate taxonomic determinations of ethnobotanically important plants. Adequately made and labeled herbarium vouchers, together with scientific determinations, provide an etic frame of reference for understanding and comparing indigenous and folk ethnobotanical knowledge systems. Chapter 8 discusses specialized collecting techniques used by ethnobotanists interested in the phytochemical aspects of economically important plant resources. Such guidelines are particularly relevant given the current interest in the value of tropical plant resources in industrial and agricultural development (Balick, 1985; Chadwick & Marsh, 1990).

Part III presents three complementary approaches to the application of quantitative techniques in ethnobotanical research.

1. Incidentally, the term *informant* has negative connotations in several countries, and its use is avoided by some in favor of terms such as *research participant* or *consultant*. In many cases, however, these alternative terms do not adequately describe the specific role of people who provide ethnobotanical or cultural information in a broad range of interviewing contexts. The term here is used in the same way that it is used in the anthropological literature.

Chapter 9 critically reviews several analytical techniques used to quantify knowledge of plant use. Chapter 10 introduces concepts and techniques from the field of human ecology that ethnobotanists wishing to quantify ethnobotanical behavior, as opposed to knowledge, might find useful. These two approaches are complemented by Chapter 11, in which ecological methods to quantify such parameters as plant resource distribution, abundance, and yields are discussed within the context of ethnobotanical research. In addition to discussing valuable techniques, Part III provides pointers to help students develop a more rigorous approach to data collection and analysis.

Part IV consists of three appendices intended to help students locate additional written resources in botany, anthropology, and ethnobotany. Specific readings in cultural, medical, and ecological anthropology, ethnoscience, linguistics, and botany, together with different approaches to ethnobotany, are included as a starting point from which readers can gain access to a broader range of written resources.

Despite, or perhaps because of, its many challenges, ethnobotany remains a fascinating and promising area of study. The increased communication between different fields and specialists, the continual development of new techniques and technologies, and continually changing ecological, social, and political contexts, demand a continuous reassessment of the assumptions, goals, and means used by ethnobotanists in the study of plant–people interactions. We hope that these guidelines will encourage and facilitate this process.

Literature Cited

Akerle, O. 1988. Medicinal plants and primary health care: An agenda for action. Fitoterapia **59:** 355–363.

Alcorn, J. B. 1984. Huastec Mayan ethnobotany. University of Texas Press, Austin.

———. 1989. Process as resource: The traditional agricultural ideology of Bora and Huastec resource management and its implications for research. Pages 63–77 *in* D. A. Posey & W. Balée, eds., Resource management in Amazonia. Advances in Economic Botany **7.** The New York Botanical Garden, Bronx.

———. 1995. The scope and aims of ethnobotany in a developing world. Pages

23–39 *in* R. E. Schultes & S. von Reis, eds., Ethnobotany. Evolution of a discipline. Dioscorides Press, Portland, Oregon.

Argueta, A., B. Torres & L. Villiers. 1982. Análisis de las categorías antropocéntricas empleadas en los estudios ethnobotánicos. Pages 32–35 *in* A. Bárcenas, A. Barrera, J. Caballero & L. Durán, eds., Memorias Simposio de Ethnobotánica, México D. F., 1978. Instituto Nacional de Antropología e Historia, México.

Atran, S. 1985. The nature of folk-botanical life forms. American Anthropologist **87**: 298–315.

Balée, W. 1994. Footprints in the forest: Ka'apor ethnobotany—The historical ecology of plant utilization by an Amazonian people. Columbia University Press, New York.

Balick, M. J. 1985. Useful plants of Amazonia: A resource of global importance. Pages 339–368 *in* G. T. Prance & T. E. Lovejoy, eds., Amazonia. Pergamon Press, New York.

———. 1990. Ethnobotany and the identification of therapeutic agents from the rainforest. Pages 21–23 *in* D. J. Chadwick & J. Marsh, eds., Bioactive compounds from plants. John Wiley & Sons, Chichester, England.

Barrera, A. 1982. La etnobotánica. Pages 6–11 *in* A. Bárcenas, A. Barrera, J. Caballero & L. Durán, eds., Memorias Simposio de Etnobotánica, México D. F., 1978. Instituto Nacional de Antropología e Historia, México.

Bellamy, R. (comp). 1993. Ethnobiology in tropical forests: Expedition field techniques. Expedition Advisory Center, Royal Geographic Society, London.

Bennett, B. C. 1992. Plants and people of the Amazonian rainforests: The role of ethnobotany in sustainable development. BioScience **42**: 599–607.

Berlin, B. 1992. Ethnobiological classification: Principles of categorization of plants and animals in traditional societies. Princeton University Press, Princeton, N.J.

———, **D. E. Breedlove & P. H. Raven.** 1974. Principles of Tzeltal plant classification: An introduction to the botanical ethnography of a Mayan-speaking people of highland Chiapas. Academic Press, New York.

Bodley, J. H. & F. C. Benson. 1979. Cultural ecology of Amazonian palms. Reports of Investigations No. 56. Laboratory of Anthropology, Washington State University, Pullman.

Boom, B. 1990. Ethics in ethnopharmacology. Pages 147–153 *in* D. A. Posey & W. L. Overal, eds., Ethnobiology: Implications and applications. Proceedings of the First International Congress of Ethnobiology, Belém, Pará, July 1988. Museu Paraense Emílio Goeldi, Belém.

Brockway, L. 1979. Science and colonial expansion: The role of the British Royal Botanic Gardens. American Ethnologist **6**: 449–465.

Browner, C. H. 1985. Criteria for selecting herbal remedies. Ethnology **24**: 13–32.

Brush, S. B. 1993. Indigenous knowledge of biological resources and intellectual property rights: The role of anthropology. American Anthropologist **95**: 653–686.

————— & D. Stabinsky (eds.). 1996. Valuing local knowledge. Indigenous people and intellectual property rights. Island Press, Washington, D.C.

Chadwick, D. J. & J. Marsh. 1990. Bioactive compounds from plants. Wiley, New York.

Conklin, H. C. 1954. The relation of Hanunóo culture to the plant world. Dissertation. Yale University, New Haven, Conn.

Cunningham, A. B. & F. T. Mbenkum. 1993. Sustainability of harvesting *Prunus africana* bark in Cameroon: A medicinal plant in international trade. People and Plants Working Paper 2. UNESCO, Paris.

Denevan, W. M. & C. Padoch (eds.). 1987. Swidden-fallow agroforestry in the Peruvian Amazon. Advances in Economic Botany 5. The New York Botanical Garden, Bronx.

Elisabetsky, E. 1986. New directions in ethnopharmacology. Journal of Ethnobiology 6(1): 121–128.

—————. 1991. Sociopolitical, economical and ethical issues in medicinal plant research. Journal of Ethnopharmacology 32: 235–239.

Ellen, R. F. & D. Reason. 1979. Classifications in their social context. Academic Press, New York.

Etkin, N. L. 1986. Multidisciplinary perspectives in the interpretation of plants used in indigenous medicine and diet. Pages 2–29 *in* N. L. Etkin, ed., Plants in indigenous medicine and diet. Redgrave, Bedford Hills, N.Y.

————— (ed.). 1992. Plants in indigenous medicine and diet: Biobehavioral approaches. Redgrave, Bedford Hills, N.Y.

—————. 1993. Anthropological methods in ethnopharmacology. Journal of Ethnopharmacology 38: 93–104.

Farnsworth, N. L., O. Akerle, A. S. Bingel, D. D. Soejarto & Z. Guo. 1985. Medicinal plants in therapy. Bulletin WHO 63: 965–981.

Ford, R. I. 1978a. Ethnobotany: Historical diversity and synthesis. Pages 33–49 *in* R. I. Ford, ed., The nature and status of ethnobotany. Anthropological Papers No. 67. Museum of Anthropology, University of Michigan, Ann Arbor.

—————. 1978b. Introduction. Pages 29–32 *in* R. I. Ford, ed., The nature and status of ethnobotany. Anthropological Papers No. 67. Museum of Anthropology, University of Michigan, Ann Arbor.

Given, D. R. & W. Harris. 1994. Techniques and methods of ethnobotany as an aid to the study, evaluation, conservation, and sustainable use of biodiversity. Commonwealth Secretariat, London.

Harris, D. R. & G. C. Hillman (eds.). 1989. Foraging and farming: The evolution of plant exploitation. Unwin Hyman, London.

Hays, T. E. 1974. Mauna: Explorations in Ndumba ethnobotany. Dissertation. University of Washington, Seattle.

Hernandez X., E. 1970. Exploración etnobotánica y su metodología. Colegio de Postgraduados, Escuela Nacional de Agricultura, Secretaría de Agricultura y Ganadería, Chapingo, México.

Hunn, E. 1982. The utilitarian factor in folk biological classification. American Anthropologist 84: 830–847.

Jain, S. K. 1989. Methods and approaches in ethnobotany. Society of Ethnobiology, Lucknow, India.

Johns, T. 1990. With bitter herbs they shall eat it: Chemical ecology and the origin of human diet and medicine. University of Arizona Press, Tucson.

Lipp, F. J. 1989. Methods for ethnopharmacological fieldwork. Journal of Ethnopharmacology **25:** 139–150.

Lewis, W. H. & M. P. F. Elvin-Lewis. 1977. Medical botany: Plants affecting man's health. Wiley, New York.

Lozoya, X. 1976. El instituto Méxicano para el estudio de las plantas medicinales. Pages 243–248 *in* X. Lozoya, ed., Estado actual del conocimiento en plantas medicinales de Méxicanas. IMEPLAM, México.

Martin, G. J. 1995. Ethnobotany: A methods manual. Chapman & Hall, New York.

Matossian, M. A. K. 1989. Poisons of the past: Molds, epidemics and history. Yale University Press, New Haven, Conn.

Moerman, D. E. 1979. Symbols and selectivity: A statistical analysis of Native American medical ethnobotany. Journal of Ethnopharmacology **1:** 111–119.

Padoch, C. 1990. Santa Rosa: The impact of the forest products trade on an Amazonian place and population. Advances in Economic Botany **8:** 151–158.

Pelto, P. J. & G. H. Pelto. 1978. Anthropological research. The structure of inquiry. Cambridge University Press, New York.

Phillips, O. L. B. & A. Gentry. 1993. The useful woody plants of Tambopata, Peru. I: Statistical hypotheses tests with a new quantitative technique. Economic Botany **47:** 15–32.

————, ————, C. Reynel, P. Wilkin, & C. Gálvez-Durand B. 1994. Quantitative ethnobotany and Amazonian conservation. Conservation Biology **8:** 225–248.

Plotkin, M. J. 1995. The importance of ethnobotany for tropical forest conservation. Pages 147–156 *in* R. E. Schultes & S. von Reis, eds., Ethnobotany. Evolution of a discipline. Dioscorides Press, Portland, Oregon.

Posey, D. A. 1982. Indigenous knowledge and development: An ideological bridge to the future. Ciencia e Cultura **35(7):** 877–894.

————. 1990. Intellectual property rights: What is the position of ethnobiology? Journal of Ethnobiology **10(1):** 93–98.

———— & W. Balée (eds.). 1989. Resource management in Amazonia: Indigenous and folk strategies. Advances in Economic Botany **7.** The New York Botanical Garden, Bronx.

Prance, G. T. 1991. What is ethnobotany today? Journal of Ethnopharmacology **32:** 209–216.

————. 1995. Ethnobotany today and in the future. Pages 60–68 *in* R. E. Schultes & S. von Reis, eds., Ethnobotany. Evolution of a discipline. Dioscorides Press, Portland, Oregon.

————, W. Balée, B. M. Boom & R. L. Carneiro. 1987. Quantitative ethnobotany and the case for conservation in Amazonia. Conservation Biology **1:** 296–310.

Redford, K. H. & C. Padoch (eds.). 1992. Conservation of neotropical forests: Working from traditional resource use. Columbia University Press, New York.

Reichel-Dolmatoff, G. 1989. Biological and social aspects of the Yuruparí complex of the Amazonian Vaupés territory. Journal of Latin American Lore **15(1):** 95–135.

Robineau, L. & D. D. Soejarto. 1996. TRAMIC: A research project on the medicinal plant resources of the Caribbean. Pages 317–325 *in* M. J. Balick, E. Elisabetsky & S. A. Laird, eds., Medicinal resources of the tropical forest. Biodiversity and its importance to human health. Columbia University Press, New York.

Salick, J. 1992. Amuesha forest use and management: An integration of indigenous forest use and natural forest management. Pages 305–332 *in* K. H. Redford & C. Padoch, eds., Conservation of neotropical forests: Working from traditional resource use. Columbia University Press, New York.

Schultes, R. E., T. Swain & T. C. Plowman. 1977. *Virola* as an oral hallucinogen among the Boras of Peru. Botanical Museum Leaflets **25:** 259–272.

——— **& S. von Reis.** 1995. Ethnobotany. Evolution of a discipline. Dioscorides Press, Portland, Oregon.

Thrupp, L. A. 1989. Legitimizing local knowledge: From displacement to empowerment for Third World people. Agriculture and Human Values **(Summer 1989):** 13–24.

Toledo, V. M. 1986. La etnobotánica en Latinoamerica: Viscitudes, contextos y desafíos. Pages 13–33 *in* IV Congreso Latinoamericano de Botánica. Simposio de Etnobotánica. Perspectivas en Latinoamerica. Serie Memorias de Eventos Cientificos Colombianos. Instituto Colombiano para el Fomento de la Educación Superior, Bogotá.

———. 1995. New paradigms for a new ethnobotany: Reflections on the case of Mexico. Pages 75–88 *in* R. E. Schultes & S. von Reis, eds., Ethnobotany. Evolution of a discipline. Dioscorides Press, Portland, Oregon.

Turner, N. J. 1988. "The importance of a rose": Evaluating the cultural significance of plants in Thompson and Lillooet Interior Salish. American Anthropologist **90:** 272–290.

Tyler, S. A. 1969. Cognitive anthropology. Holt, Rinehart & Winston, New York.

Weller, S. C. & A. K. Romney. 1988. Systematic data collection. Sage Publications, Newbury Park, California.

Werner, O. & G. M. Schoepfle. 1987. Systematic fieldwork. Ethnographic analysis and data management. Sage Publications, Newbury Park, California.

Wickens, G. E. 1990. What is economic botany? Economic Botany **44:** 12–28.

I
Conducting Ethnobotanical Research

Introduction

Although every field situation presents unique logistical, methodological, theoretical, and ethical challenges, certain themes and questions tend recur in most studies. Thus, all researchers require consent from one or more of the parties involved in the study, though obtaining consent may involve anything from getting official permits to gaining community approval. People with whom ethnobotanists interact are usually from a different social class, community, language group, or nationality, so researchers must find a way to explain their objectives and needs accurately yet meaningfully. All ethnobotanists deal with cultural knowledge that may be regarded as the intellectual property of individuals, communities, and societies, and all ethnobotanists need to consider the foreseeable impacts of their research on the communities in which they work. Finally, every fieldworker needs to find and develop viable relationships with informants and must have the necessary skills to record cultural data accurately.

In Chapter 1, Alexiades presents general guidelines to help researchers successfully initiate and conduct ethnobotanical fieldwork. This process includes obtaining necessary permits, communicating effectively with communities and informants, and selecting and compensating informants. The subject of adequate compensation and intellectual property rights is further developed and explored by Cunningham, who reviews different approaches to this complex and important topic. Social scientists have had a long-standing involvement with the ethical implications of their research, but discussions of ethics have not entered the mainstream ethnobotanical discourse until recently. Although ethnobotany has always played a direct and indirect role in the commoditization of indigenous knowledge and natural resources, recent advances in the fields of plant breeding, molecular biology, and phytochemical screening have highlighted this role and the need to consider what constitutes adequate amounts and forms of compensation. These two chapters emphasize the degree to which perceptions and approaches to this important topic are constantly changing and illustrate how the dialectics of the interaction between the researcher and informants constantly challenge the researcher and his or her assumptions.

In Chapter 3, Alexiades discusses concepts and techniques central to the collection of cultural data. Adequate cross-cultural communication and interviewing skills are essential if many human–plant interactions, including plant use knowledge, are to be recorded in a meaningful way. Alexiades's paper emphasizes how an adequate understanding of the "native's point of view"—how informants structure and organize their experience—is necessary to ensure that the research results are not simply artifacts of the research design. Such an emic understanding complements the etic approaches subsequently discussed in Part III.

I

Protocol for Conducting Ethnobotanical Research in the Tropics

Miguel N. Alexiades
Institute of Economic Botany,
The New York Botanical Garden

Introduction

Ethnobotanical fieldwork requires researchers to collaborate with government officials, scientists, and local people. The following guidelines are intended to help ethnobotanists establish field re-

Selected Guidelines for Ethnobotanical Research: A Field Manual, 5–18
Edited by Miguel N. Alexiades
© 1996 The New York Botanical Garden

search projects and develop fruitful professional relationships. Additional guidelines for research conduct and protocol are provided elsewhere (Boom, 1990; Colvin, 1992; Mori & Holm-Nielsen, 1981; Pearson, 1985; Society for Economic Botany, 1994). The "Declaration of Belém," signed during the First International Congress of Ethnobiology in Belém, Brazil, in July 1988, provides broad guidelines delimiting the relationship between indigenous peoples, scientists, and policy makers (Posey, 1990). Finally, Kloppenburg and Balick (1996) and Reid et al. (1993) provide a general discussion and protocol guidelines related to prospecting for biodiversity. These should be of interest to all ethnobotanists, particularly those involved in commercially based research.

Obtaining Permits and Permission

In most countries, researchers must obtain official permits before conducting any research, particularly when collecting plant specimens or when working in indigenous communities.[1] Often, they must deal with national as well as regional government and nongovernment agencies. Although the specific procedure and requirements for obtaining research and collecting permits vary, researchers are frequently asked to (1) be affiliated with a research institution or university; (2) develop formal collaboration with an institution of the host country, in the case of foreign scientists; and (3) deposit duplicates of botanical collections and reports in host-country institutions. This process often takes several months, and researchers are advised to find out what the specific requirements are and apply for permits well ahead of time. Bypassing this bureaucratic procedure is illegal and unethical: it gives fieldworkers a poor reputation and jeopardizes the possibility for future research.

Fieldworkers should also obtain permission from community members before initiating research activities. The correct protocol may vary, but it usually involves establishing contact with

[1] Daly and Beck (Chapter 8, this volume) discuss permits for collecting bulk plant samples for phytochemical study. Ethnobotanists working for the private sector and "biodiversity prospectors," collecting material for phytochemical and genetic study, should also consult Janzen et al., 1993, for protocol guidelines related to conducting commercial research and obtaining collecting permits.

the representative agency for the group under study. In peasant and indigenous communities, this step often entails contacting the community leader first and then participating in a community or leaders' meeting. When working in indigenous communities, researchers should also obtain the approval of the regional or national indigenous federation with which the community is affiliated. At the very least, these organizations should be informed of any research activities. Obtaining approval often involves signing a contract or agreement (often referred to as a *convenio* in Latin America) that formalizes the conditions under which the research will be conducted, including mechanisms of compensation and exchange of information, services, and resources. Clearly, a community might feel that the research is not in their best interests and decide that they do not want it to take place in their area. Besides obtaining formal permission from the relevant authorities and community representatives, plant collectors should also seek permission of individual owners when collecting near houses, in gardens, or on cultivated land.

During public and private discussions with informants, researchers need to clearly and honestly discuss their goals, methodology, and the foreseeable implications and consequences of the research. An open mind and flexibility are necessary in all negotiations. Local communities or organizations may already have experience with other researchers, or they may have suggestions that could improve the research or add to its value to the local community.

Communicating the aims and consequences of the research is a complex matter and may be harder than it sounds. Informants may have a totally different frame of reference and experience than the researcher, and they may not be familiar with many of the institutions and concepts central to the research question. As a result, informants may not comprehend the full implications of their participation in the research. For example, informants may not be aware of how publication of their intellectual property could compromise their future ability to obtain financial compensation for it. This difficulty is compounded by the fact that no researcher can predict the full long-term impact and consequences of the research. Thus, all researchers can do is discuss these issues honestly and in a way that is meaningful to informants. Clearly, as the relationship between the researcher and his or her informants changes over time, so do the issues that need to be discussed and the ability to discuss them (Mann, 1981). Discussion of the research aims and possible consequences should not be a single event that takes place at the outset of the research; it should be an ongoing process that begins with negotiations and remains an important part of the relationship between the researcher and the people with whom she or he works.

Research Design and Participatory Approaches

The nature and context of ethnobotanical fieldwork are such that researchers often will work with human groups and areas confronting tremendous social, health, environmental, and political problems. The degree of the researcher's involvement in and commitment to these issues will, of course, vary according to the specific circumstances, as well as his or her personal and professional goals. In any case, ethnobotanists have a responsibility to the people with whom they work (Society for Economic Botany, 1994), and this should at least include an attempt to incorporate, at some level, local needs into the research design (Greaves, 1996). Doing so may include discussing research goals during early negotiations with participants and modifying these if necessary.

The subject matter of ethnobotany, the relationship between people and their botanical resources, is ideally suited to applied and participatory research. As indigenous and folk societies become increasingly organized and vocal with respect to their roles in conservation and resource management (see, e.g., Bodmer et al., 1990; COICA, 1990; Schwartzman, 1989; Varese, 1996), the role of ethnobotanists and the nature of their interactions with their informants are likely to change (Toledo, 1995). An appropriate question in this regard is, At what point should research participants be considered research colleagues or consultants (see e.g., Berlin, 1984), not only with respect to research design and implementation but also with respect to publication of the research results?

Although the words *participation* and *participatory* are used very frequently in development and applied research, they can have different meanings, depending on the decision-making capacity awarded to local people. In participatory rural appraisal discourse, participation is seen as a mechanism whereby local people gain control of the research process—thus breaking the relationship of dependency that characterizes most interactions between technocrats and local peoples—and ultimately achieve greater self-reliance and self-determination (Jordan, 1989). In this context, the researcher is seen as a facilitator who helps channel people's knowledge and creativity into the research process. Manuals for participatory research training emphasize the need to be flexible and open-minded and to develop the ability to lead but not direct (Davis-Case, 1989). Recent advances in the application of participatory methodologies in agriculture, community forestry, and health care (Davis-Case, 1989; Donnelly-Roark, 1987; Jordan, 1989; National Environment Secretariat et al., 1991) suggest that this research technique has considerable potential in ethnobotanical research.

Other Protocols

Ethnobotanists have a specific moral and professional responsibility to establish and maintain links and collaborative exchanges with host-country institutions and colleagues. Collaborative ventures are valuable to all parties and demonstrate a commitment to helping advance the field in the host country. Mori and Holm-Nielsen (1981) recommend that, before arriving in the host country, visiting scientists who wish to collect plant specimens let resident botanists know the planned dates of visit and determine beforehand who will number which collections, how many duplicates will be collected, and how these will be distributed. Duplicates should always be deposited in at least one herbarium in the host country: in fact, this is often stipulated as a condition for obtaining collecting permits.

Fieldworkers should also ensure that the results of their research, including publications, are available to colleagues in host countries. Every effort should be made to disseminate their research findings by giving seminars and publishing research findings in the host country's journals and language. Whenever possible, researchers should also help students from host countries acquire field experience and develop research skills (Pearson, 1985).

Ethnobotanists should be mindful of how their work and actions may be interpreted by the people with whom they work or live. For example, indigenous and folk cultures often have particular regard or respect for some plant species considered to be endowed with powerful medicinal, religious, or magical properties. Collecting these plants without asking permission or without conducting certain rituals beforehand may be offensive and compromise people's faith in the ethnobotanist's suitability as a recipient of their knowledge. In general, fieldworkers should conduct their affairs with cultural sensitivity, based on knowledge and respect of local norms and expectations for conduct and behavior. Clearly, any acts that may be considered offensive or that are apt to be misinterpreted should be avoided.

Selecting Informants

Ethnobotanical fieldwork is a cooperative venture between the researcher and local informants. It is very important to establish good rapport with at least several people in the community before selecting and working with informants. It is particularly important that researchers communicate their objectives and expectations clearly and honestly before initiating the fieldwork and explain all aspects of the research, including the need to collect plant specimens. Researchers should expect informants to have concerns about the ethnobotanist's motives, and these need to be repeatedly addressed, formally and informally, in public and private meetings.

Selecting and obtaining informants requires social skills, which in turn should build upon a sense of cultural sensitivity and respect. It often behooves the worker to be aware of the "political map" of the community: researchers need to consider how their actions may be interpreted in light of community and gender relations, and they should attempt to foresee and understand the impact of their actions on community dynamics. Therefore, it is most helpful if workers can spend some time in the community before beginning "active" research.

A common mistake when interacting with unfamiliar cultures is assuming that one is dealing with a culturally homogeneous group and that culture is a monolithic entity that can be studied through interaction with the most "authoritative" informant

(Kemp & Ellen, 1984). Although much ethnobotanical information is widely shared, there is often a considerable degree of intracultural variation and specialization in ethnobotanical knowledge, even within small communities (Berlin, 1992; Boster, 1986; Posey, 1992). These factors need to be taken into account when making decisions as to the sample size and selection of informants (see also Alexiades, Chapter 3, this volume). In-depth analysis of ethnobotanical knowledge and processes necessarily restricts the sample size and the degree to which the study can be considered representative. Ultimately, the goals of the study should help determine an adequate compromise between its depth and breadth.

Compensation of Informants and Communities and Intellectual Property Rights

Fieldworkers should be prepared to discuss and negotiate the ways in which their research might benefit the people involved in the study, either as individuals, as a community, or both. Any commercial objectives and expected profits should be clearly stated at the beginning of the negotiations with the community and informants. Benefits are not necessarily limited to direct monetary compensation, though this often may be a just and important component (see Cunningham, Chapter 2, this volume). The question of compensation is particularly complex in ethnobotany because ethnobotany has direct and indirect links with specific commercial and political interests (Balandrin et al., 1985; Brockway, 1979; Juma, 1989; Kloppenburg, 1988). As academic-corporate interactions become an increasingly important part of research funding, the need to develop adequate means of compensation has become increasingly apparent (Boom, 1990). Even purely academic research cannot fully disassociate itself from commercial implications. For one thing, academic researchers have no control over the use and applications of their published data, and disclosure of ethnobotanical knowledge could in some cases jeopardize the ability of folk societies to obtain adequate compensation for the subsequent commercial use of their intellectual property.

Compensation, including what constitutes a fair amount, who should receive it, and the mechanisms by which it should be channeled, is a complex and controversial subject (Barton, 1994; Greaves, 1994; Patel, 1996; Cunningham, Chapter 2, this volume). Clearly, the form and degree of "fair" compensation will depend on the characteristics and duration of the study. Informants and communities can be compensated formally and informally. Formal compensation is usually discussed and agreed upon *before* fieldwork commences. In many cases, formal compensation is most appropriate when dealing with institutions and communities, leaving informal compensation more to interactions with individuals. Formal compensation mechanisms are particularly important and relevant when dealing with projects that have commercial objectives or when any income is expected from the sale of articles, books, or slides. Just because compensation is "informal," it does not mean that it is less important. Furthermore, formal and informal compensation mechanisms are usually not mutually exclusive but rather are complementary and used to different degrees depending on the context.

King et al. (1996) outline some ways in which one pharmaceutical company provides immediate reciprocal benefits to indigenous people for use of their knowledge. These include supporting efforts to secure land rights and providing or contributing to community development programs. The characteristics of ethnobotanical information are ideally suited for this form of exchange. For example, researchers could help produce a health worker's manual on medicinal plants when the study has been completed[2] or facilitate reforestation or the planting of kitchen gardens with local food, medicinal, or industrial species. Other possibilities include helping to build a school or purchasing books or educational materials for the local school. Compensation can also be provided through services offered to the community, by facilitating transfer of appropriate technology, or by training local personnel.

Where commercial payoffs are expected, a number of workers have stressed the need for such legal compensatory mechanisms

[2]Some ethnobotanists feel that researchers need to distinguish between the descriptive and prescriptive functions of ethnobotany and that scientists should advocate the use of medicinal plants only after these have been screened for safety and efficacy.

as advance payments and royalties and patents, stipulated as part of contractual agreements (Laird, 1993; Cunningham, Chapter 2, this volume). Posey (1991) reviews some of the international institutional and legal mechanisms currently available for the implementation of intellectual property rights. Brush (1993) discusses the origins and characteristics of the concept of intellectual property rights, together with its applications and limitations as a means of determining compensation. Additional treatments relating to intellectual property rights and local knowledge have been published by Brush and Stabinsky (1996), Greaves (1994), and Swanson (1995).

In most cases, it is also appropriate to compensate individual informants directly for their time. Direct monetary payment in exchange for time or information is one possibility, but in some circumstances it may raise difficulties. Some informants may be tempted to provide "information" as a way of acquiring compensation, especially if it is provided in proportion to the amount of time given or information provided. In some cases, restricting compensation to informants alone may create resentment and conflicts with other members of the community. These difficulties do not negate the value of direct payments as an option for compensation; rather they serve to forewarn researchers of potential pitfalls.

A second possibility is to compensate informants through gifts and services. In some cases, this manner of compensation may be more akin to local modes of exchange and reciprocity. For example, the researcher might bring a small gift or some food when visiting an informant at home. Gifts (possibly including cash) may be given at the start or end of the research. The researcher might also help a family pay for the necessary arrangements for a traditional feast or ritual. The researcher's assistance is almost always sought during emergencies or on special occasions, such as accidents, weddings, births, burials, and so on. There usually is no lack of opportunities to show gratitude and to reciprocate.

The general idea in these latter cases is to move away from compensation as a transaction toward compensation as a vehicle for building a relationship based on reciprocity and trust. At times, both modes of exchange will be necessary and just, depending on the duration and characteristics of the study and the expectations of informants. In general, the longer the fieldworker is able to stay in an area, the more he or she can participate in nontransactional exchanges, building a sense of friendship and trust with informants.

Researchers should be wary of making promises they cannot keep and of raising unrealistic expectations. Doing either may not only create disappointment and resentment but may also jeopardize future research plans for the researcher and others. Similarly, researchers should be prepared to deal with the fact that individuals, communities, or institutions may overestimate the researcher's ability to provide compensation.

As with other ethical issues, compensation has to be dealt with on a case-by-case basis. Appropriate arrangements in one setting may be inappropriate in another. What is clear, however, is that (1) compensation is necessary; (2) local expectations must be taken into account during formal and informal negotiations; (3) fieldworkers have the professional responsibility to try to ensure that compensation is just and equitably distributed; and (4) compensation is provided in a form that is truly beneficial.

Recognition of intellectual property rights has consequences beyond compensation. For example, researchers should obtain permission from informants before disclosing their intellectual property, and the possibility of coauthorship and an active participation in the process of designing and publishing research results should be considered. At the very least, informants should be acknowledged in publications (unless they prefer to remain anonymous).

Timing

Timing is a critical consideration in any study as it places very clear constraints on the scope and objectives of the study, including the relationship that can be developed with informants and local participants. Clearly, the longer the time in the field, the greater the opportunity to build rapport and collect more data. Most ethnobotanical studies require an initial investment of time in reconnaissance and rapport-building; the duration of this phase clearly depends on the specific circumstances. Researchers should also bear in mind the relationship between the timing of their fieldwork and seasonal changes. During certain times of the year, for example, people might be more available to participate in the study than they would be during others. More importantly, perhaps, timing should account for seasonal variations in how people interact with plants. The researcher might remain in the field for a year, or return to the field for a shorter time in different seasons. Seasonality may also have important logistical implications; for example, many tropical areas remain inaccessible or hard to reach during monsoon or the rainy season.

Acknowledgments

I am grateful to William Balée, Michael Balick, Bradley Bennett, Scott Mori, Christine Padoch, Daniela Peluso, Charles Peters, Oliver Phillips, Jennie Wood Sheldon, and two anonymous reviewers for their helpful comments on earlier drafts, and to May Ebihara for her helpful introduction to several of the topics discussed. Any errors of interpretation are my own. The Institute of Economic Botany of The New York Botanical Garden supported the preparation of this paper.

Literature Cited

Balandrin, M. F., J. A. Klocke, E. S. Wurtele & W. H. Bollinger. 1985. Natural plant chemicals: Sources of industrial and medicinal materials. Science **228**: 1154–1160.

Barton, J. H. 1994. Ethnobotany and intellectual property rights. Pages 214–227 in G. T. Prance, D. J. Chadwick & J. Marsh, eds., Ethnobotany and the search for new drugs. Ciba Foundation Symposium 185. John Wiley, Chichester, England.

Berlin, B. 1984. Contributions of Native American collectors to the ethnobotany of the Neotropics. Advances in Economic Botany **1**: 24–33.

———. 1992. Ethnobiological classification: Principles of categorization of plants and animals in traditional societies. Princeton University Press, Princeton, N.J.

Bodmer, R., J. Penn, T. G. Fang & L. Moya. 1990. Managing programs and protected areas: The case of the Reserva Comunal Tamshiyacu-Tahuayo, Peru. Parks **1**: 21–25.

Boom, B. 1990. Ethics in ethnopharmacology. Pages 147–153 in D. A. Posey & W. L. Overal, eds., Ethnobiology: Implications and applications. Proceedings of the First International Congress of Ethnobiology, Belém, Pará, July 1988. Museu Paraense Emílio Goeldi, Belém.

Boster, J. S. 1986. Exchange of varieties and information between Aguaruna manioc cultivators. American Anthropologist **88**: 428–436.

Brockway, L. 1979. Science and colonial expansion: The role of the British Royal Botanic Gardens. American Ethnologist **6**: 449–465.

Brush, S. B. 1993. Indigenous knowledge of biological resources and intellectual property rights: The role of anthropology. American Anthropologist **95**: 653–686.

────── & **D. Stabinsky (eds.)** 1996. Valuing local knowledge. Indigenous people and intellectual property rights. Island Press, Washington, D.C.

Coordinadora de las Organizaciones Indígenas de la Cuenca Amazónica (COICA). 1990. The Iquitos Declaration. Cultural Survival Quarterly **14(4)**: 82.

Colvin, J. G. 1992. A Code of ethics for research in the Third World. Conservation Biology **6**: 309–311.

Davis-Case, D. 1989. Participatory monitoring and evaluation: A field manual. Community Forestry Unit of the Food and Agriculture Organization, Rome.

Donnelly-Roark, P. 1987. New participatory frameworks for the design and management of sustainable water supply and sanitation projects. WASH Technical Report No. 52. PROWESS Report No. 50. Promotion of the Role of Women in Water and Environmental Sanitation Services/United Nations Development Program, Washington, D.C.

Greaves, T. (ed.). 1994. Intellectual property rights for indigenous people: A sourcebook. Society for Applied Anthropology, Oklahoma City.

──────. 1996. Tribal rights. Pages 25–40 in S. B. Brush & D. Stabinsky, eds., Valuing local knowledge. Indigenous people and intellectual property rights. Island Press, Washington, D.C.

Janzen, D. H., W. Hallwachs, R. Gámez, A. Sittenfield & J. Jímenez. 1993. Research management policies: Permits for collecting and research in the tropics. Pages 131–157 in W. V. Reid, S. A. Laird, C. A. Meyer, R. Gámez, A. Sittenfield, D. H. Janzen, M. A. Gollin & C. Juma, eds., Biodiversity prospecting: Using genetic resources for sustainable development. World Resources Institute, Washington, D.C.

Jordan, F. (comp.). 1989. Capacitación y participación campesina. Instrumentos metodológicos y medios. Instituto Interamericano de Cooperación para la Agricultura, San José, Costa Rica.

Juma, C. 1989. The gene hunters: Biotechnology and the scramble for seeds. Princeton University Press, Princeton, N.J.

Kemp, J. H. & R. F. Ellen. 1984. Producing data: Informal interviewing. Pages 229–236 in R. F. Ellen, ed., Ethnographic research: A guide to general conduct. Academic Press, New York.

King, S. R., T. J. Carlson & K. Moran. 1996. Biological diversity, indigenous knowledge, drug discovery, and intellectual property rights. Pages 167–185 in S. B. Brush & D. Stabinsky, eds., Valuing local knowledge. Indigenous people and intellectual property rights. Island Press, Washington, D.C.

Kloppenburg, J. R., Jr. (ed). 1988. Seeds and sovereignty: The use and control of plant genetic resources. Duke University Press, Durham, N.C.

———— & M. J. Balick. 1996. Property rights and genetic resources: A framework for analysis. Pages 142–173 in M. J. Balick, E. Elisabetsky & S. A. Laird, eds., Medicinal resources of the tropical forest. Biodiversity and its importance to human health. Columbia University Press, New York.

Laird, S. A. 1993. Contracts for biodiversity prospecting. Pages 99–130 in W. V. Reid, S. A. Laird, C. A. Meyer, R. Gámez, A. Sittenfield, D. H. Janzen, M. A. Gollin & C. Juma, eds., Biodiversity prospecting: Using genetic resources for sustainable development. World Resources Institute, Washington, D.C.

Mann, B. J. 1981. The ethics of fieldwork in an urban bar. Pages 95–109 in M. A. Rynkiewich & J. P. Spradley, eds., Ethics and anthropological dilemmas in fieldwork. Robert E. Krieger Publishing, Malabar, Fla.

Mori, S. A. & L. B. Holm-Nielsen. 1981. Recommendations for botanists visiting neotropical countries. Taxon **30(1):** 87–89.

National Environment Secretariat, Egerton University & Clark University. 1991. Participatory rural appraisal handbook: Conducting PRA's in Kenya. Center for International Development and Environment, World Resources Institute, Washington, D.C.

Patel, S. J. 1996. Can the intellectual property rights system serve the interests of indigenous knowledge? Pages 305–322 in S. B. Brush & D. Stabinsky, eds., Valuing local knowledge. Indigenous people and intellectual property rights. Island Press, Washington, D.C.

Pearson, D. L. 1985. United States biologists in foreign countries: The new ugly American? Bulletin of the Ecological Society of America **66:** 333–337.

Posey, D. A. 1990. Introduction to ethnobiology: Its implications and applications. Pages 1–8 in D. A. Posey & W. L. Overal, eds., Ethnobiology: Implications and applications. Proceedings of the First International Congress of Ethnobiology, Belém, Pará, July 1988. Museu Paraense Emílio Goeldi, Belém.

————. 1991. Effecting international change. Cultural Survival Quarterly **15(3):** 29–35.

————. 1992. Interpreting and applying the "reality" of indigenous concepts: What is necessary to learn from the natives? Pages 21–34 in K. H. Redford & C. Padoch, eds., Conservation of neotropical forests: Working from traditional resource use. Columbia University Press, New York.

Reid, W. V., S. A. Laird, C. A. Meyer, R. Gámez, A. Sittenfield, D. H. Janzen, M. A. Gollin & C. Juma (eds.). 1993. Biodiversity prospecting: Using genetic resources for sustainable development. World Resources Institute, Washington, D.C.

Schwartzman, N. B. 1989. Extractive reserves: The rubber tapper's strategy for sustainable use of the Amazon rainforest. Pages 150–165 in J. O. Browder, ed., Fragile lands of Latin America: Strategies for sustainable development. Westview Press, Boulder, Colo.

Society for Economic Botany. 1994. Guidelines of professional ethics of the Society for Economic Botany. Society for Economic Botany Newsletter **7 (Spring 1994):** 10.

Swanson, T. M., ed. 1995. Intellectual property rights and biodiversity conservation: An interdisciplinary analysis of the values of medicinal plants. Cambridge University Press, New York.

Toledo, V. M. 1995. New paradigms for a new ethnobotany. Reflections on the case of Mexico. Pages 75–88 *in* R. E. Schultes & S. von Reis, eds., Ethnobotany. Evolution of a discipline. Dioscorides Press, Portland, Oregon.

Varese, S. 1996. The new environmentalist movement of Latin American indigenous people. Pages 122–142 *in* S. B. Brush & D. Stabinsky, eds., Valuing local knowledge. Indigenous people and intellectual property rights. Island Press, Washington, D.C.

2

Professional Ethics and Ethnobotanical Research

Anthony B. Cunningham
WWF/UNESCO/Kew People and Plants Initiative

Introduction
Parallel Debates: Crops, Novel Compounds, and Ornamental Plants
Views of Indigenous Peoples
Levels of Sensitivity: Species and Local Knowledge
Plants, Power, and Specialist Knowledge
Commercial and Academic Interests: The Example of New Natural Products
Key Questions and Actions
  Who should be informed about planned research work?
  How do local people's schedules fit in with the timing of research-related meetings?
  How can we deal with unrecorded knowledge?
  How should research participants be paid?
  How should the intellectual contribution of research partners be recognised?
  How do your research objectives fit in with national or local priorities?
  Can you assist in developing a national research capability?
  Can you comply with requirements for confidentiality or anonymity from the people or community you are working with, or, if you are doing commercially relevant work, with the confidentiality requirements of the sponsoring organisation?
  Do the research requirements comply with national legislation and ethical guidelines of the professional society to which you belong?

Selected Guidelines for Ethnobotanical Research: A Field Manual, 19–51
Edited by Miguel N. Alexiades
© 1996 The New York Botanical Garden

Will resource overexploitation result from your research work?
If you undertake work with commercial intent, what contractual arrangements exist to ensure an equitable distribution of benefits should a product with commercial value be discovered?
If you undertake work with commercial intent, what information should
you give regarding the possible commercial benefits of the research?
What form should benefits take, and how should they be disbursed?

Introduction

Ethnobiologists and anthropologists play an important role in recording traditional knowledge before it disappears through cultural assimilation or environmental change. These records are
important for their cultural value as well as for the ecological
insights traditional knowledge provides into ecosystem functioning and resource management. Indigenous knowledge is also a
key to plants useful to a much wider sector of society, such as
potential new crop plants or pharmaceutical drugs, insecticides,
and other industrial products.

In recent years, with political, economic, and technological
change, researchers have become increasingly aware of the socioeconomic and political context of their work. One aspect is the
need to conduct research *with* local people rather than purely *for*
or *about* them. This research approach has been facilitated in the
ethnobotanical and agricultural research fields through a range
of participatory methods that bring together the technical and
theoretical strengths of Western science with indigenous knowledge (FAO, 1990; Pimbert, 1994; Poffenberger et al. 1992). Researchers and local people participating in research are in very
different positions of economic and political power, however.
Ethnobiologists, economic botanists, and anthropologists often
are from urban, academic backgrounds. The people who are the
richest source of indigenous knowledge are usually from rural
areas with little access to formal education. They also have the
least economic or political power within their national govern-

ments. Hunter-gatherer societies, for example, universally recognised for their acute observation and ecological insights, are extremely vulnerable to material and political marginalisation and displacement by pastoralists and agriculturalists. Most researchers have trained in an academic environment, where the ideal of free flow of information is a very important principle; yet confidentiality of some forms of information also needs to be taken into account, either for the safety or privacy of local research participants[1] or for commercial reasons.

Funding to many academic institutions has declined with a decrease in government subsidies. Consequently, research organisations have become increasingly competitive for funds. Most researchers are familiar with the adage "publish or perish." In the commercial world, however, researchers are familiar with the importance of *not* making information publicly available before it is patented. Questions of funding and timing of publication are very relevant to economic and ethnobotanists as well as to other researchers working on developing new natural products. The need for consideration of ethics in ethnobotany has been stimulated by new developments in biotechnology and genetic engineering and the close relationship between industry and researchers based at universities and botanical gardens. Researchers often find themselves in a position of brokers between two worlds, particularly where specialist knowledge is concerned. In some cases, researchers are able to obtain and record knowledge from traditional specialists such as herbalists, diviners, beekeepers, or master fishermen only after establishing credibility and a position of trust with them and the local community. Detailed information usually can be obtained only after an extended period, as some information may be shielded from nonspecialists even within that community by initiation rites and taboos. This position of trust should extend to the way that the various forms of information derived from ethnobotanical work are used and stored. These forms include the intellectual information from local specialist plant users as well as genetic and chemical information from plants.

[1] The term *research participants* is used here as a deliberate alternative to *informants* or *collaborators,* terms that can have very negative connotations in many conflict-ridden societies. As Preston-Whyte (1987) pointed out, it also more accurately describes the increasing role of local people in a participatory research process.

Parallel Debates: Crops, Novel Compounds, and Ornamental Plants

Three main categories of economically important plants have been the focus of what are often parallel debates pertinent to professional ethics, traditional knowledge, and plant conservation. First, in the early 1980s international debate focused on local knowledge, farmers' rights, and equity in the distribution and controls of crop plant genetic resources (Busch et al., 1991; Crucible Group, 1994; Mooney, 1983). At the time, these issues drew heated argument from some prominent botanists (Arnold et al., 1986), but they were largely resolved when the Commission on Plant Genetic Resources (PGR) revised the FAO Undertaking for Plant Genetic Resources to recognise both plant breeders' rights and farmers' rights (World Resources Institute, 1992). Second, the more recent debate is related to development of new natural products, particularly medicinal plants but also soil fungi and marine and terrestrial animals (Hamburger et al., 1991). Third, debate continues about the distribution of economic benefits from plant species with horticultural potential. All of these debates hinge on professional approaches related to:

- recognition of the intellectual contribution made by farmers or traditional specialists (such as herbalists, beekeepers, and master fishermen) to the development or identification of crop landraces and useful wild plants;
- equitable distribution of benefits from the use of crops, wild plants, or their genetic or chemical structures to assist people and plant conservation in the region of origin, attributable to the wide recognition that economics can play a role in justifying conservation as a form of land use and to the need to maximise "value-adding" from resource harvesting, including the collection of plant genetic material and chemical structures;
- the degree of the research organisation's commitment to technology transfer, infrastructure development, training programs and the local government's support of development of crop varieties, new natural products, and horticultural exports.

These debates question the view that indigenous knowledge and chemical and genetic resources should be freely available as global common property. The Convention on Biological Diversity (signed in Rio de Janeiro, June 1992), for example, has led to greater awareness of the need for national policies relating to research protocols and collection of biological material. Growth of the fields of genetic engineering and biotechnology has stimulated interest in potential industrial products from plants. As a consequence, professional organisations involved in ethnobiological and natural-products chemistry research have taken important steps to clarify their codes of ethics (Table I). These codes emphasise the need to develop equitable partnerships for "capturing" and effectively dispersing benefits from these resources, rather than considering them a global commons.

Responsibility for resolving these issues rests not only on the professional ethics of researchers but also on the policies of governments, research funding organisations, and indigenous peoples. In 1989, for example, UNESCO adopted a resolution on the Safeguarding of Traditional Culture and Folklore (UNESCO, 1989). In 1991, the FAO Commission on Plant Genetic Resources produced a draft International Code of Conduct for Plant Germplasm Collecting and Transfer, which was updated in 1993 (FAO, 1991, 1993). This was a voluntary code of conduct addressed to international and national agricultural research institutions, recognising farmers' rights and setting up guidelines for the exchange of germplasm. National guidelines on access to biological resources are also being developed, for example, in Australia (Department of the Prime Minister and Cabinet, 1994).

Views of Indigenous Peoples

At a meeting in Penang, Malaysia, in 1992, representatives of indigenous peoples drafted a Charter of the Indigenous-Tribal Peoples of the Tropical Forests in response to the cultural threats of encroachment and forest destruction. This charter called for several actions relevant to research ethics and biodiversity conservation, including:

• the right to be informed, consulted and above all to participate in decision-making on legislation and policies, and in the formulation,

Table I. Summary of issues relating to ethics, ethnobotany, and new natural products development covered in resolutions or statements of various organisations

Organisation	Equitable partnerships	Training/ technology transfer	Health in developing countries	Sustainable resource use	Survey species & traditional knowledge	National sovereignty
International Society of Ethnobiology (1988)	*	*	*	*	*	*
International Society of Chemical Ecology (1989)	*			*	*	
Society of Economic Botany (1990)	*	*			*	
NIH/NCI workshop of Drug Development, Biological Diversity & Economic Growth (1991)	*	*	*	*	*	
Global Biodiversity Strategy (WRI/IUCN and UNEP) (1992)	*	*	*	*	*	*
ASOMPS (Asian Symposium for Medicinal Plants, Spices and Other Natural Products) (1992)	*	*	*	*	*	*
American Society of Pharmacognosy (1992)	*	*	*	*	*	*

Source: From Cunningham, 1993.

implementation or evaluation of any development project, whether local, national or international, private or of the state, that may affect our futures directly or indirectly;

- promotion of the health systems of the indigenous peoples, including the revalidation of traditional medicine, and the promotion of programmes of modern medicine and primary health care. Such programmes should allow us to have control over them, providing suitable training to allow us to manage them ourselves;
- promotion of alternative fiscal policies that permit us to develop our community economies and develop mechanisms to establish fair prices for the products of our forests;
- programmes related to biodiversity must respect the collective right of our peoples to cultural and intellectual property, genetic resources, gene banks, biotechnology and knowledge of biodiversity; this should include our participation in the management of any such project on our territories, as well as control of benefits that derive from them;
- since we highly value our traditional technologies and believe that our biotechnologies can make important contributions to humanity, including "developed" countries, we demand guaranteed rights to our intellectual property, and control over the development and manipulation of this knowledge;
- all investigations in our territories should be carried out with our consent and under joint control and guidance according to mutual agreement; including provision for training, publication and support for indigenous institutions necessary to achieve such control.

More recently, the Mataatua Declaration made recommendations on the cultural and intellectual property rights of indigenous peoples (United Nations, 1993) and Posey et al. (1994) reviewed existing international legal and ethical provisions that can strengthen the protection of community intellectual property and traditional resource rights.

Levels of Sensitivity: Species and Local Knowledge

Economic botanists and ethnobotanists are concerned both with people and with biodiversity conservation. Specialist traditional knowledge of endemic species, genera, or families represents both the category of indigenous knowledge and the category of cultural sensitivity and conservation concerns (Table II). This category is equally the focus of academic and commercial interests.

An example is the University of Illinois/National Cancer Institute (NCI) drug discovery programme, which focuses on "endemic species that, based on local knowledge, are employed for the treatment of cancer, for wound healing, for improvement of health (excluding foods) or which seem to have an immunostimulant effect" (Farnsworth, pers. comm., 1988). Owing to the symbolic and often religious nature of traditional medicines and specialist knowledge of their use, particular sensitivity needs to apply in how this category is recognised or applied (Table II). Similarly, particular care has to be taken with the way that endemic species or plants vulnerable to overexploitation are collected.

Plants, Power, and Specialist Knowledge

From the viewpoint of many traditional healers in southern Africa, certain specialist knowledge should certainly not be made public. Even within their own societies, much of this knowledge is kept private through ritual and taboo. Similarly, in Australia, Martu aboriginal people chose to censor research information on medicinal plants after it had been recorded, as it was considered to be very powerful (Walsh, pers. comm., 1992). First, it was requested that information on a particular medicinal plant species not be disclosed, as the Martu people did not want outsiders using the plant or making money from the medicine. Second, there was concern that inexperienced people, for their own safety, should not be exposed to powerful medicines. In another case, although no restrictions were placed on information on widespread useful species, Martu people requested that vernacular names of localised plants used by the Martu not be published. If "whitefellas" wanted to find these plants in Martu country, they would have to work through the Martu people. Similar views are expressed by indigenous peoples in other parts of the world out of concern for what often are perceived not as global common property but as locally important resources. Many traditional healers recognise that their knowledge has a wider value and want part of the benefits arising from its use. Even in remote areas of Africa, Latin America, and Asia, there is widespread awareness amongst indigenous peoples of the linkage between traditional and pharmaceutical medicines. There is also awareness of the interest of industrial companies in developing new drugs from plants or of commercial values attached to crop plant genetic resources. Over a decade ago, the Organisation of African Unity (OAU) went so far as to urge secrecy in herbal medicine research in order to prevent multinational companies from developing new drugs and selling them back to the country of origin at high prices (Hanlon, 1979).

If neither indigenous knowledge nor biodiversity can be conserved, then the result is a double tragedy where all will be losers. This is one of the reasons why mutually acceptable research agreements need to be developed through consultation and acceptable guidelines for researchers. In Panama, for example, Kuna people working together with researchers have established guidelines for this cooperative work (Box 1). A further example is the health care and forest conservation initiative developed with local communities, university scientists, and doctors through a WWF conservation programme for Manongarivo Special Reserve, Madagascar (Quansah, 1994), which includes the development of a role for an ethical pharmaceutical company.

The library at the Australian Institute for Aboriginal and Torres Strait Islander Studies provides another example of a practical approach that combines controlled access to information with the ethical responsibility of researchers. As with all libraries, it is an invaluable store and source of openly available information. In addition, the Institute also recognises that culturally important information cannot be seen in isolation and that the Institute has a responsibility to Aboriginal people as well as to scholarship (Anonymous, 1990). For this reason, certain material is not freely available, nor is it openly displayed. Even when published, conditions on the sale of the publication are that:

> any purchaser or subsequent reader shall abide by a specific condition placed on secret/sacred material in this book by Aboriginal men and/or women. All knowledge relating to those rituals (sites) is normally confined to the men and/or women who have been inducted into them. Many Aboriginal people are eager to have this material recorded and published as a matter of permanent record. Because much of this material is secret/sacred it may cause great distress if it is discussed with any Aborigine before it has been established that he/she has the correct standing in his/her society and is willing to participate in discussion.

(text continued on page 30)

Table II. Sensitivity levels, by category of knowledge, species and distribution, habitat, and harvested products and genetic or chemical values by demand:supply ratio. Greatest ethical challenges are centered on the smallest categories: unpublished specialist indigenous knowledge of endemic species and where harvesting of popular but scarce, slow-growing resources in high-diversity habitat is occurring or is proposed, particularly where the requirement is for bark, roots, hardwoods, or whole plants. This last is often the category of greatest commercial or academic interest.

Sensitivity level	Indigenous knowledge	Plant taxa	Harvesting	Genetic or chemical value	Demand:supply ratio
Low	Common knowledge in "local public domain" (e.g., bush foods, fuelwoods, building materials). Some traditional ecological knowledge (TEK)	Cosmopolitan species, global distribution (e.g., *Typha latifolia*, *Phragmites*)	Widespread, large area, high species biomass production, high reproductive rate, low habitat specificity. Often resilient, weedy, or wetland species	Plant taxa with no crop relatives and no forage values or ethnobotanical use	Low

Medium	Some knowledge limits by gender, age, or skill (e.g., household herbal remedies, fish toxins). Detailed TEK.	Pantropical/subtropical distribution, or pantemperate distribution.	Common; rapid growth rates; fruit, flower, or leaf harvesting nondestructive.	Wild relatives of non-commercial crops, low forage value, low potential source of new natural products	Medium
High	Symbolic/ritual uses, specialist medicinal uses	Endemic or near endemic taxa; e.g., *Ancistrocladus korupensis* (Nigeria, Cameroon), *Calophyllum lanigerum* (Sarawak), *Conospermum* (Australia)	Highly habitat-specific. Use of bark, root, hardwood, stem, or whole plant. Slow-growing/reproducing species. High habitat diversity (e.g., tropical forests). National parks or culturally sensitive sites	Taxa with many active ingredients (e.g., Solanaceae, Menispermaceae, Euphorbiaceae), oils or resins. Forage or crop relatives (e.g., *Coffea*, *Zea*).	High

Box 1: Case Study: Project for the Study of the Management of Wildlife Areas of Kuna Yala (Pemasky), Panama

Established by the Kuna Indian people of Panama in 1983 with substantial international funding and the support of the Smithsonian Tropical Research Institute (STRI), this project has aimed at the management of a 60,000-ha forest reserve within Kuna land. Research is carried out by non-Kuna scientists working with Kuna assistants. The STRI, based in Panama City, have played an important role in ensuring that research permission is first obtained by scientists and that research reports, photographs, and biological specimens are left with the PEMASKY project.

In 1988, the Kuna produced a 26-page manual, *Research Program: Scientific Monitoring and Cooperation,* with guidelines and information for visiting researchers. This gives detailed information on how to apply for permission to visit the area, which sites are off limits, and how various activities such as plant collecting and marking of animals are viewed.

The manual sets guidelines without being overly bureaucratic, stating, for example, that:

> All researchers should consider the incorporation of Kuna co-researchers, assistants, guides and informants, with the objective of training Kuna scientists, and achieving a transference of knowledge and technologies. . . . The principal researcher will consider paying his assistants and other (local) informants.

Guidelines for feedback from research carried out in the area are also specified, stating that:

> Each researcher is asked to send two (2) copies of his publication on the research carried out. . . . If possible, summaries of abstracts should be translated into Spanish.

This approach offers an opportunity for benefits to both researchers and research partners.

Source: From Chapin, 1991.

Commercial and Academic Interests: The Example of New Natural Products

One consequence of limits on funding to academic institutions has been the need to source funding through contracts with industrial companies. An example is the funding to universities and botanical gardens as part of new drug discovery programmes (see

Table III). The value of the contracts awarded by the United States Developmental Therapeutics Program (DTP) of the U.S. National Cancer Institute is a measure of the interest in new natural products drugs discovery. In 1986, three five-year contracts worth US$2.7 million were awarded. In 1991, these were renewed, with the three contracts valued at US$3.8 million (NCI, 1992). The three contractors, Missouri Botanical Garden, The New York Botanical Garden, and the University of Illinois at Chicago, which subcontracts to the Arnold Arboretum (Harvard University) and Bishop Museum, Honolulu, are all based in the United States. Collections are focused respectively on Africa and Madagascar, Central and South America, and Southeast Asia. These collaborative programmes have resulted in the discovery of some interesting new compounds, including those with anti-HIV properties from the Ancistrocladaceae, Combretaceae, Euphorbiaceae, and Piperaceae (Kashman et al., 1992). This involves the collection of biological samples on commercial contract, sometimes guided by indigenous knowledge or sometimes on a taxonomic basis.

Commercial interest in collecting biological material or recording specialist traditional knowledge raises questions about the academic tradition of free flow of information. In the development of natural products and in "chemical prospecting," biological samples and research knowledge are not freely disseminated into the public domain. Research information, such as data on chemical structures of active ingredients, needs to be kept confidential if it is to be patented before it is made public. By contrast, biological samples and indigenous knowledge have been considered a "global commons" in the past. This attitude, often unknowingly fostered by researchers involved in collecting or developing these resources, raises important questions relating to intellectual property rights (Posey, 1991). It also devalues the phytochemical resources that can help justify maintenance of species-rich vegetation through value-added effects as part of new natural products development. Collecting programmes for new natural products development therefore raise two topical issues of research ethics. The first involves the ethical and legal arguments relating to intellectual property rights and the use of indigenous knowledge. The second involves conservation requirements to avoid natural resource exploitation and assist in the

(text continued on page 34)

Table III. Pharmaceutical companies and research organisations involved in screening plants for new natural products, showing focus of programme and sources of supply

Organisation	Status of plant screening programme	Plant material supplied by	Region of origin
American National Cancer Institute (NCI)	Large-scale screening of plants and marine organisms	Missouri Botanical Garden, New York Botanical Garden, University of Illinois, private contractors	Africa, Madagascar, Central and South America, South-east Asia, Australia
Bristol-Meyers	None at present; evaluating whether to include plants or not; developed taxol from Pacific yew (*Taxus*)	Not applicable	Taxol material from USA
Glaxo	Natural products discovery department; many therapeutic areas	Commercial and academic institutions; Royal Botanic Gardens, Kew	South America, Africa

Merck, Sharp Dohme Research Laboratories	Marine organisms, plants, and microorganisms	New York Botanical Garden; work with IN-Bio, Costa Rica	South America
Monsanto/Searle	Microorganisms and plants	Missouri Botanical Garden	North America
Shaman Pharmaceuticals	Plants, on basis of ethnobotanical information	Individuals, institutions, and government departments	Tropical South America, Africa, Southeast Asia
SmithKline Beecham	Marine organisms, plants, and microorganisms	Biotics Ltd., UVA, private individuals, and own collectors	Malaysia, Micronesia

Source: From Findeisen, 1991, and miscellaneous other sources.

justification of conservation as a form of land use through channeling value-added benefits from natural products development back to the source country (Eisner & Beiring, 1994; Reid et al., 1993).

Under the "common property system," which considers genes or chemical structures from biological material as the common heritage of mankind and an asset belonging to nobody, there is little incentive to conserve either species or habitats. "Chemical prospecting," as Eisner (1990) termed it, can be a powerful ally compatible with biological conservation if a fair share of benefits reverts to the region of origin. Research agreements are one of the means through which this goal will occur (Janzen et al., 1993; Reid et al., 1993), possibly through linkages to a Biotic Exploration Fund, as suggested by Eisner and Beiring (1994) (Figure 1). This approach is not without its critics, however. Internationally known U.S. research chemist Carl Djerassi (1992) questions why developing countries should be paid any royalty at all from new drug development:

> After a few million dollars are spent to bring such a product to the regulatory approval stage, should royalties be paid to Third World countries where a few grams or even kilograms of the original plant were collected? Or for a product originally derived from a marine organism collected within the frequently claimed (and disputed) 200-mile territorial limit of certain countries? Suppose the plant came from Switzerland? Should the royalties be paid to a Swiss canton by Lilly, Glaxo, or Ciba-Geigy? If we wish to contribute to the economic well-being of a Third World country—and I am all for it—let us do it on more logical grounds.

Two things need to be borne in mind, however. First, the "few grams or even kilograms" of plant material might be extremely valuable if it provides initial chemical leads to new compounds and new industrial products. Second, "chemical prospectors" are unlikely to "prospect" for new drugs from plants in Switzerland but would collect biological samples from vegetation with high numbers of endemic taxa. Switzerland, France, Germany, and the United Kingdom may be rich in technology and per capita income, but they are a poor source of biological material. One need only compare the number of endemic plant species in Switzerland (1 endemic species), Germany (73 endemic species), or the United Kingdom (16 endemic species) with those

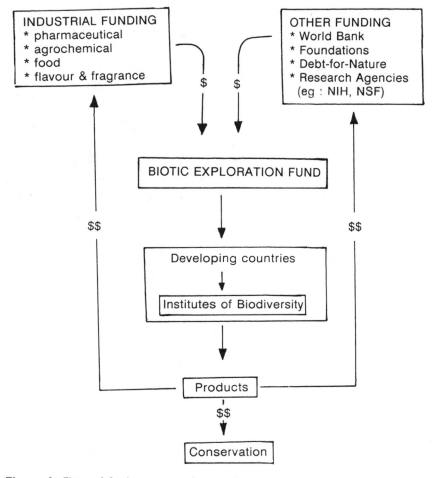

Figure I. Flow of funding sources for the Biotic Exploration Fund proposed by Eisner and Beiring (1994).

in countries where chemical prospecting is taking place, such as Mexico (3376 endemic species) or the Amazonian region (25,000–30,000 endemics). It will be clear why the tropics are a focus of most chemical prospecting. As Janzen and collaborators (1993) pointed out:

> With so much at stake, the traditional view of the raw materials used in research on tropical wildland biodiversity—whether data, samples or specimens—as "free goods" must be abandoned, and . . . guidelines [must be] expanded to cover ecotourists, school groups, private collectors, taxonomists, collectors for pharmaceutical and biotechnol-

ogy research, wildland managers, national developers and many other kinds of users.

Key Questions and Actions

Both researchers and ethical companies have a key role to play as "honest brokers." There is also a great need for coordinated effort at international, regional, national, and local levels. The ethics of interaction between researchers and traditional communities is the responsibility of the individual researcher, the professional organisation or society to which the researcher belongs, and the national governments. The 1992 Manila Declaration concerning the ethical use of Asian Biological Resources is a notable exception to most policy statements. In addition to setting out policy guidelines, the Manila Declaration set out a clear and concise code of ethics for foreign collectors of biological samples and a set of minimum contract guidelines (Cruz et al., 1992) (Box 2).

The following key questions, pitfalls, and principles may be useful as guidelines for ethnobotanical research.

Who should be informed about planned research work?

There is a need to discuss the objectives, methods, and likely outcomes of research with local communities (see Alexiades, Chapter 1, this volume). Researchers should fully inform research participants (e.g., traditional specialists) and members of relevant local organisations (e.g., national herbaria) of the objectives, commercial aspects, and possibly results of the research. Confidential information and requests for anonymity of research participants need to be respected. Arrangements for equitable compensation for assistance by individuals and fair royalty payments to the relevant national or regional organisation need to be made.

ACTION   National requirements for plant collecting, including collection with local counterparts, and international phytosanitary and conservation legislation (such as Convention on International Trade in Endangered Species of Wild Fauna and Flora [CITES] regulations) need to be followed (see Box 2).

Box 2: Contract Guidelines

These guidelines were developed at the ASOMPS VII meeting in Manila, Philippines (Cruz et al., 1992). They recognize that there is considerable variation in the levels of technical expertise for new natural products in the Asian region. In the absence of sufficient expertise to undertake screening within the country, it has been suggested that in the short term, efficient development of new natural products may involve sharing of biological resources and technology between developed countries and countries of origin of biological material. The following minimum standards were suggested in the Manila Declaration:

(i) the amount of material collected for initial screening should not normally exceed 500 grams (dry weight) unless specific permission is obtained;

(ii) payment should include all handling expenses;

(iii) where screening is carried out with the aid of a partner organisation in the developed world, a minimum of 60% of any income arising from the supply of extracts to commercial organisations should be returned to the appropriate country organisation;

(iv) the country organisation should receive a minimum of 51% of any royalties arising from external collaboration that result in marketable products. As fair royalties are normally 3%–5%, the national organisation would be expected to receive a minimum royalty of 1.5%–2.5%;

(v) the country organisation should not sign agreements that give exclusive, indefinite rights to any external party. Exclusivity should be limited to no more than a two year period;

(vi) complete evaluation results should be reported to the supplying country within 6–9 months;

(v) if there is a threat of destructive harvesting, costs of sustainable harvesting or development of alternative supplies must be borne by external organisation;

(vi) the contribution of research participants should be recognised through coauthorship of publications (unless anonymity has been requested);

(vii) where possible, screening of extracts should be carried out in the country of origin, and assistance be provided to develop this expertise wherever practicable.

How do local people's schedules fit in with the timing of research-related meetings?

Consultation and participatory research are important processes. Researchers need to be aware, however, of the time they can take from other community activities. Research surveys, for example, are time-consuming, particularly when groups are involved. Appropriate means of consultation, research methods, and timing of meetings can help minimise the consumption of a community's time.

ACTION   Researchers need to ask the local people: How busy are you? Do you really want to be involved at every phase of research, considering that you may have crops to weed, livestock to tend, families to feed, fuel and water to collect, or other daily tasks? What are the key issues that you want to discuss and resolve?

How can we deal with unrecorded knowledge?

Inventories of published specialist data already in the public domain, such as the NAPRALERT and PHARMEL databases, provide an extremely useful source of information, not only on what is published but also on what information is not yet published. Care needs to be taken when recording unpublished information in deciding how it should be stored or made available to other people. Research participants should understand the research objectives and have the right to decide what is made public and what is not. Previously unrecorded knowledge is best approached through participatory research at a local community level. As Johns (pers. comm., 1993) pointed out, this approach provides a means for people whose knowledge is sought (1) to come to an agreement with the researcher on research goals and feedback information to the community, (2) to set any limitations on the type of information to be disseminated, and (3) to come to an agreement on reasonable compensation, or (4) to refuse to participate.

How should research participants be paid?

Compensation needs to be made when extensive use is made of local people's time and expertise. Fair compensation will vary in form from area to area and on the type of research, and it is best to discuss what constitutes fair payment with local counterparts or research organisations so that neither local people nor future research is disadvantaged. Purchase of information is often not practical, as it could tempt respondents into giving nonsense an-

swers for material gain. On the other hand, man\
community, either in group discussions or as indivi\
time and effort assisting researchers. This time woul\
be spent in a range of important activities, whether \
fields, looking after children, gathering fuelwood, or se\
ducc in markets.

How should the intellectual contribution of research partners be recognised?

Few ethnobotanists would dispute the intellectual contribution made by skilled research participants. Although research partici-pants may not be competing in the "publish or perish" research world, recognition of this intellectual contribution can have posi-tive benefits through reaffirming the global and local value of traditional knowledge. When and how this intellectual contribu-tion is acknowledged, or whether it is desirable at all, need to be discussed with research participants. In some cases, depending on the content and context of the research, anonymity may be re-quested and preferable to public recognition. In other cases, indi-viduals or even an entire community may require acknowledg-ment (such as Baker & Mutitjulu Community, 1992).

ACTION   Before starting research, researchers need to set aside suf-ficient time and be well prepared for discussion with community leaders, representative group(s) of local people, and specialist plant users to discuss and reach consensus on these issues. De-pending on the society and circumstances, some of these ques-tions may be discussed individually or in groups. Unless ano-nymity is required, appropriate recognition should be given to assistance received from host-country research participants. In addition, the Melaka Accord (ASOMPS, 1994) has suggested that:

- journal editors, peer reviewers, and professional societies, when re-viewing manuscripts, should try to ensure that . . . recognition is given;
- in countries where permit infrastructure exists, all researchers and particularly collectors should formally acknowledge permit approval (by citation of permit number or equivalent) in reports and manu-scripts, with copies sent to the permit authorities.

How do your research objectives fit in with national or local priorities?

Ethnobotanical research results are often of value to local people and applicable to rural development issues such as health, energy needs, housing, land-use planning, and natural resource management. The same applies to new natural products development, which should be relevant to the health and conservation needs of developing countries. Priorities within developing countries may be very different from those in the temperate zone; there may be a greater need for drugs to treat parasitic, respiratory, or diarrheal diseases that are major health problems in developing countries or to assist in evaluating traditional remedies.

In many parts of the world, both saving species and preserving indigenous knowledge are urgent issues at a time when both trained researchers and research funds are scarce. Under these circumstances, waste of funds due to duplicated or poorly directed research needs to be avoided. This can only be done after thorough preparation that increases awareness of related research work that has already taken place.

ACTION   Develop research objectives with local research organisations and local communities in response to perceived needs and problems and provide feedback on the progress and results to research participants. To avoid duplication of research, communicate with other researchers working in the region.

Can you assist in developing a national research capability?

Visiting researchers should be aware of the research capability within the country or region and that they can contribute to local research capability through funding and training of graduate or undergraduate students. Both visiting and local researchers as well as local traditional experts can benefit from collaborative work through sharing of expertise, whether in collecting or evaluating samples. In the case of natural products development, for example, the "Letter of Intent" of the National Cancer Institute makes provision for the involvement by the company or universities involved in training local staff (e.g., senior technicians, scientists, or postgraduate students) from the country or region of origin of the plant. This develops local expertise for natural products identification and development. It may be more effective and preferable, however, to have an arrangement whereby a visiting researcher trains local people in the country of origin. One visiting research specialist would be able to train many local researchers, with less chance of the trainees' being tempted to leave the country for more lucrative employment elsewhere.

ACTION Wherever possible, visiting researchers should involve local students, researchers, and traditional experts as research participants through seminars and training. Local researchers must call for political commitment to training, technology transfer, and support for local scientists and research organisations.

Can you comply with requirements for confidentiality or anonymity from the people or community you are working with, or, if you are doing commercially relevant work, with the confidentiality requirements of the sponsoring organisation?

Researchers working with traditional specialists such as herbalists are in a position of trust, and, as Padoch and Boom (1990: 10–11) pointed out, researchers:

- should not "trick" research participants or informants into revealing "secret information";
- should respect the rights of research partners to anonymity and privacy when they are requested;
- will disclose to their sponsors their need to comply with the ethical guidelines of their professional society, including the stipulation that those studied will be fully informed concerning the objectives, including commercial ones, and possible results of the research.

For companies, there is a need for exclusivity and confidentiality of test results by all parties for a set period of time until a patent application has been filed on any potentially valuable active ingredients. Researchers must be clear that this confidentiality requirement will not overturn other ethical principles. They may also be able to arrange for a limited embargo period, after which the information would become more freely available. Once such agreements have been made, it is important to ensure that all parties adhere to them and that samples from the same species are not sold to other companies during this period.

ACTION Respect the right of privacy of research partners. If commercial collecting is undertaken, clearly discuss with local counterparts the requirement for exclusive rights for a set period as a condition of contract work.

Do the research requirements comply with national legislation and ethical guidelines of the professional society to which you belong?

In addition to legal requirements, there are guidelines set by the professional societies to which many researchers belong. These include the need for openness vis-à-vis relevant government organisations, NGOs (nongovernment organizations), and community leaders regarding the aims and methods of the research, its sponsors, and whether the research is conducted with commercial intent. Researchers also need to ensure that their research will not jeopardise any of the people with whom they are working, whether during the research or after it has been completed— this includes the right of any research partner to anonymity. By the same token, researchers should ensure that no species or habitat is endangered as a result of their work. Again this may require a situation of anonymity of either the location or even identity of rare or localised species that may be threatened by commercial interests. This anonymity may be either temporary or permanently restricted to certain organisations or individuals. All of these issues challenge the admirable academic ideal of free flow of information. This challenge is increasingly taking place as previously "neutral" academic or nonprofit organisations are funded through linkage to commercial organisations.

National legislation governs the collecting and export of biological specimens and, in cases of foreign researchers, stresses the need to collect with local counterparts. Adequately documented and mounted specimens need to be left with appropriate national or regional institutions. In the past and at present collectors are exporting raw material (or encouraging poorly informed local collectors to do so) with either no voucher specimens deposited with national herbaria or research institutions, or, if vouchers are deposited, they are poor and inadequately documented specimens.

Action Adequately annotated and preserved specimens of biological material need to be lodged with appropriate national institutions. If visiting foreign collectors are doing the research, they should comply with national regulations on the collection and export of biological material and cover the costs of collecting with at least one local counterpart. If adequate legislation does not yet exist, then national researchers need to press for adequate legislation to control the collection and export of biological material with advice from appropriate professional organisations.

Will resource overexploitation result from your research work?

More and more ethnobotanists are becoming involved in resource-management recommendations on use of nontimber forest products by local people. There are two reasons. First, there is a renewed interest in promoting the use of nontimber forest products as a development alternative, either for local marketing or for export (Clay, 1991; Phillips, 1993). Second, local people's uses of plants are increasingly being taken into account in resource-sharing arrangements in or around national parks and protected areas. This aim generates some local income and employment, but it may lead to destructive, species-selective harvesting of wild populations. A focus on wild harvest may also result in little or no attempt to develop or cultivate alternative supplies, with the resultant possibility that a local multiple-use resource is depleted. It is important that poorly founded recommendations do not contribute to this problem.

Large-scale harvesting by foreign contract collectors also takes place. Collectors usually take initial samples for screening, ranging from 100 g to a few kilograms of plant material. Extracts are prepared and screened for activity against conditions such as cancer or HIV infection or as anti-inflammatories. In many cases, re-collections are required, to collect sufficient material for isolation and identification of the active compound(s). In this case, bulk samples may be collected by professional contract plant collectors. Under these circumstances it may be necessary to protect the identity of the plant material until a contract has been signed. This protection could be accomplished through coding extracts as, in some cases, bulk collection is made possible because of reference to herbarium material collected by academic researchers. This coding can ensure that unethical companies will not overexploit the species concerned. It may also facilitate an agreement for cultivation, which ensures a more sustainable source of income and employment than will a "boom and bust" approach, in which wild stocks are overexploited.

ACTION Researchers should avoid making poorly founded recommendations for multiple-use or commercial harvesting. Contracts for new natural products development should also specify that if enough sustainably harvested material is not available from wild populations, then cultivation needs to be supported to prevent overexploitation of wild stocks. Cultivation should preferably take place in the country of origin and may include development and selection of fast-growing cultivars that provide high yields of the active ingredient(s). This procedure should take place as a matter of course if endemic or near-endemic species are involved.

If you undertake work with commercial intent, what contractual arrangements exist to ensure an equitable distribution of benefits should a product with commercial value be discovered?

In the past, and in some cases at present, collecting took place in an uncontrolled manner, and sample material was taken (usually) to Europe, Japan, or North America for analysis. Patenting and drug development often took place without even the knowledge of people in the country or region of origin, with no recompense for use of regional natural resources, often without any contractual obligation. Local professionals (e.g., botanists, foresters) were also paid individually to collect sample material for industrial companies, which did not inform them of the implications and paid them fees that had little relation to the potential value of the resource. Many developing countries pay these staff relatively low salaries, and "hard currency" is hard to get, so it is quite understandable that local professionals were willing to collect specimens. Nevertheless the practice has important implications for regional development and conservation. No measures were set up for a percentage of profits from new natural products development to be channeled back to the region of origin should that economic potential be realised. Detailed guidelines aimed at resolving this situation have been recommended by Cruz et al. (1992) (Box 2) and through WWF-International (Cunningham, 1993).

Action If no research contract exists or if it is inadequate, tell the organisation involved so that they can review their position. If changes are inadequate, then work with more suitable organisations who have already established their credibility. Supply agreements should be made with organisations rather than individuals. Contracts should include provision for equitable distribution of benefits to research partners in the country or region of origin and should aim for minimum standards similar to those established in Box 2.

If you undertake work with commercial intent, what information should you give regarding the possible commercial benefits of the research?

Only a small percentage of active ingredients from plants are finally developed and marketed, with a "hit" rate between 1:6,000 and 1:10,000. Over a 25-year period, for example, Merck, Sharp and Dohme found only five compounds that either directly or with some chemical modification have become marketable drugs (Schweitzer et al., 1991). Development costs are high; patent rights are generally up to 20 years, but with 10–15 years of this time taken up in drug development before marketing can take place. Tyler (1986) quoted the cost of new drug development as

US$50–100 million per new product. Today, this figure is closer to US$200 million per product.

Action Ensure that unrealistic expectations do not develop amongst participants in developing countries regarding benefits from potentially important industrial products.

What form should benefits take, and how should they be disbursed?

Feedback of research information to rural communities, planners, and politicians can have a positive influence on land-use issues crucial to local people. In many cases, it is necessary to return research data in a more usable form than tables, lists, and figures in scientific jargon. Data on nutritional values of local foods or on toxic or medicinal plants can be developed into useful health-care booklets for use by community health-care workers. Recommendations on under- or overexploited wild plants can facilitate resource-sharing arrangements with national parks or provision of alternatives through agroforestry initiatives.

A complex question is how to deal with economic benefits that may arise from royalties on sales of natural products. Even when the problems surrounding the capture of benefits through legal agreements are solved, the problem of how to distribute the benefits remains. Anders (1989), for example, documented the social problems of alcoholism, social inequalities, and tensions that developed as money flowed into traditional societies. A further problem, as recognised by Schweitzer et al. (1991), is to establish intermediate forms of compensation and incentives that bridge the 10–15 years that it may take to develop a marketable drug.

Decisions on how to disburse these funds will vary with local circumstances and need to be developed through discussions with research partners and participants. They may vary, for example, from funds to institutions (e.g., universities, herbaria, botanical gardens, traditional healers' associations) to funds for conservation projects, education bursaries, or legal resources funds. In general, however, many traditional societies would benefit if sustained rural primary health-care schemes were coordinated with traditional practitioners. Skilled individuals and communities themselves can also benefit from public recognition of the value of their customary knowledge. This benefit is particularly important in countries where an emphasis on the "superiority" of urban-industrial society has given rise to a stigma that denies or hides the value of customary knowledge held by rural peoples with no formal education.

Misappropriation of funds has been a common feature of many government and nongovernment organisations, and safeguards would have to be built against it. Income from patent rights is unlikely to accrue to a specific community (as plant uses are often known through much of the range of a plant species) unless a highly localised, endemic species is involved. One possibility is to relocate benefits on a bioregional basis rather than on the basis of political boundaries. Another would be to base distribution of benefits on cultural boundaries, which, like species distributions, cross political boundaries. A regional fund, perhaps administered by an appropriate nongovernment organisation through community leaders, may be another way of disbursing funds.

Action If a new product is developed, patent or legal contract rights should ensure adequate and mutually acceptable return of benefits to the region of origin (see Box 2). A central principle should be that these benefits are distributed as locally to the source area as possible rather than into anonymous government coffers. Benefits may be in forms other than monetary benefits. Decisions on how to disburse these benefits will vary with local circumstances and need to be developed through discussions with research partners and participants.

Conclusion

Research in the ethnobotanical field has developed rapidly in the past decade. It is also taking place at a challenging time of new technological developments, a renewed search for new natural products, and a time of crisis at the culture–nature interface. A result has been a variety of ethical dilemmas for researchers. New ones will undoubtedly arise. Few individuals can be expected to be experts in all components of the cross-disciplinary research that ethnobotany represents in anthropology, chemistry, ecology, taxonomy, linguistics, or legal process. A strong case can therefore be made for multidisciplinary teams of researchers, ranging from local traditional experts to anthropologists, ecologists, chemists, taxonomists, economists, and statisticians. It is expected that mistakes, ethical or otherwise, will be made through lack·of awareness. What is clear, however, is the need to avoid unprofessional research that jeopardises local people, biological conservation, and opportunities for researchers in the future.

Acknowledgments

This chapter summarises approaches outlined in a more detailed document (Cunningham, 1993), which is available from WWF-International, Gland, Switzerland, and had benefited from comments of G. Anderson, J. Armstrong, J. Ayafor, J. Cannon, M. Cele, T. Johns, A. Hamilton, D. Harder, N. Marchant, M. Martin-Smith, M. Pimbert, D. Posey, N. Quansah, F. Walsh, P. Waterman, and colleagues at the ASOMPS VII and International Society for Ethnobiology (ISE) meetings. Financial support from WWF-International and the WWF/UNESCO/Kew People and Plants Initiative is gratefully acknowledged.

Personal Communications

N. R. Farnsworth. 1988. University of Illinois at Chicago, College of Pharmacy, 833 Wood Street, Chicago, Ill., 60612, USA.

T. Johns. 1993. School of Dietetics and Human Nutrition, McDonald College of McGill University, 21 111 Lakeshore Road, Ste. Anne de Bellvue, H9X1 CO, Canada.

F. Walsh. 1992. Department of Botany, University of Western Australia, Nedlands, 6009, Western Australia.

Literature Cited

Anders, J. C. 1989. Social and economic consequences of federal Indian policy: A case study of the Alaskan natives. Economic Development and Cultural Change **37:** 285–303.

Anonymous. 1990. Australian Institute of Aboriginal and Torres Strait Islander Studies. Research grants: Information and conditions for applicants. Union Off-set Company, Canberra.

Arnold, M. H., D. Astley, E. A. Bell, J. K. A. Bleasdale, A. H. Bunting, J. Burley, J. A. Callow, J. P. Cooper, P. R. Day, R. H. Ellis, B. V. Ford-Lloyd, R. J. Giles, J. G. Hawkes, J. D. Hayes, G. G. Henshaw, J. Heslop-Harrison, V. H. Heywood, N. L. Innes, M. T. Jackson, G. Jenkins, M. J. Lawrence, R. N. Lester, P. Matthews, P. M. Mumford, E. H. Roberts, N. W. Simmonds, J. Smartt, R. D. Smith, B. Tyler, R. Watkins, T. C. Whitmore & L. A. Withers. 1986. Plant gene conservation: Letter to *Nature* objecting to P. Mooney's views on crop plant genetic resources. Nature **319**: 615.

Asian Symposium on Medicinal Plants, Spices, and Other Natural Products (ASOMPS). 1994. The Melaka Accord towards the development of legislation to protect biodiversity. Asian Symposium on Medicinal Plants, Spices and Other Natural Products. Department of Chemistry, University Pertanian Malaysia, Serdand Selangor, Malaysia.

Baker, L. M. & Mutitjulu Community. 1992. Comparing two views of the landscape: Aboriginal ecological knowledge and modern scientific knowledge. Rangeland Journal **14(2)**: 174–189.

Busch, L., W. B. Lacy, J. Burkhardt & L. R. Lacy. 1991. Plants, power, and profit: Social, economic, and ethical consequences of the new biotechnologies. Basil Blackwell, Oxford.

Chapin, M. 1991. How the Kuna keep the scientists in line. Cultural Survival Quarterly **15(3)**: 17.

Charter of the Indigenous-Tribal Peoples of the Tropical Forests. 1992. Statement of the International Alliance of the Indigenous-Tribal Peoples of the Tropical Forests. Penang, Malaysia.

Clay, J. 1991. Some general principles and strategies for developing markets in North America and Europe for non-timber forest products. Pages 302–309 *in* M. Plotkin & L. Famolare, eds., Sustainable marketing of rain forest products. Island Press, Washington, D.C.

Crucible Group. 1994. People, plants and patents: The impact of intellectual property on trade, plant biodiversity, and rural society. International Development Research Centre, Ottawa.

Cruz, L. J., G. P. Concepcion, A. S. Mendigo & B. Q. Guevara. 1992. Seventh Asian Symposium on Medicinal Plants, Spices and Other Natural Products. APO Production Unit, Manila, Philippines.

Cunningham, A. B. 1993. Ethics, ethnobiological research and biodiversity. Research paper, WWF-International, Gland, Switzerland.

Department of the Prime Minister and Cabinet. 1994. Access to Australia's biological resources: A discussion paper. Australian Government Publishing Service, Canberra.

Djerassi, C. 1992. Drugs from Third World plants: The future. Science **258**: 203.

Eisner, T. 1990. Prospecting for nature's riches. Chemoecology **1**: 38–40.

—— & E. A. Beiring. 1994. Biotic exploration fund—Protecting biodiversity through chemical prospecting. BioScience **44**: 95–98.

FAO. 1990. The community's toolbox: The idea, methods and tools for participatory assessment, monitoring and evaluation in community forestry.

Community Forestry Field Manual 2. Food and Agriculture Organisation of the United Nations, Rome.

———. 1991. Draft international code of conduct for plant germplasm collecting and transfer. Commission on Plant Genetic Resources, Fourth Session, 15–19 April 1991. Food and Agriculture Organisation of the United Nations, Rome.

———. 1993. Agreed text on draft international code of conduct for plant germplasm collecting and transfer. Commission on Plant Genetic Resources, 103rd Session, 14–25 June 1993. Food and Agricultural Organisation of the United Nations, Rome.

Findeisen, C. 1991. Natural products research and the potential role of the pharmaceutical industry in tropical forest conservation. Periwinkle Project, Rainforest Alliance, New York.

Hamburger, M., A. Marston & K. Hostettmann. 1991. Search for new drugs of plant origin. Advances in Drug Research **20:** 167–215.

Hanlon, J. 1979. When the scientist meets the medicine men. Nature **279:** 284–285.

Janzen, D. H., W. Hallwachs, R. Gámez, A. Sittenfield & J. Jimenez. 1993. Research management policies: Permits for collecting and research in the tropics. Pages 131–157 *in* W. V. Reid, A. Sittenfield, S. A. Laird, D. H. Janzen, C. A. Meyer, M. A. Gollin, R. Gámez, & C. Juma, eds., Biodiversity prospecting: Using genetic resources for sustainable development. World Resources Institute, Washington, D.C.

Kashman, Y., K. R. Gustafson, R. W. Fuller, J. H. Cardellina, J. B. MacMahon, M. J. Currens, R. W. Buckheit, S. H. Hughes, G. M. Cragg & M. R. Boyd. 1992. The Calanolides, a novel HIV-inhibitory class of coumarin derivates from the tropical rainforest tree, *Calophyllum lanigerum.* Journal of Medicinal Chemistry **35:** 2735–2743.

Mooney, P. R. 1983. The law of the seed. Development Dialogue **1–2:** 1–172.

National Cancer Institute (NCI). 1992. Development Therapeutics Program, Division of Cancer Treatment, Letter of Intent. Frederick, Maryland.

Padoch, C. & B. Boom. 1990. Professional ethics in economic botany: A preliminary draft of guidelines. Society for Economic Botany Newsletter **4 (September 1990):** 10 11.

Phillips, O. 1993. The potential for harvesting fruits in tropical rainforests: New data from Amazonian Peru. Biodiversity and Conservation **2:** 18–38.

Pimbert, M. 1994. The need for another research paradigm. Seedling **11:** 20–25.

Poffenberger, M., B. McGean, N. H. Ravindranath & M. Gadgil. 1992. Field methods manual: Diagnostic tools for supporting joint forest management systems. Vol. 1. Joint Forest Management Programme, New Delhi.

Posey, D. A. 1991. Effecting international change. Cultural Survival Quarterly **15(3):** 29–35.

———, **A. Argumento, E. da Costa e Silva, G. Duthfield, K. Plenderleith & J. Freidman.** 1994. Indigenous peoples, traditional technologies and equitable sharing: International instruments for the protection of community intellectual property and traditional resource rights. Draft document. Oxford Centre for Environment, Ethics and Society, Oxford University.

Preston-Whyte, E. M. 1987. Research ethics in the social sciences. *In* J. Mouton & D. Joubert, eds., Knowledge and method in the human sciences. Human Sciences Research Council, Pretoria.

Quansah, N. 1994. Biocultural diversity and integrated health care in Madagascar. Nature & Resources **30**: 18–22.

Reid, W. V., A. Sittenfield, S. A. Laird, D. H. Janzen, C. A. Meyer, M. A. Gollin, R. Gámez & C. Juma (eds.). 1993. Biodiversity prospecting: Using genetic resources for sustainable development. World Resources Institute, Washington D.C.

Schweitzer, J., F. G. Handley, J. Edwards, W. F. Harris, M. R. Grever, S. A. Schepartz, G. Cragg, K. Snader & A. Bhat. 1991. Commentary: Summary of the workshop on drug development, biological diversity and economic growth. Journal of the National Cancer Institute **83**: 1294–1298.

Tyler, V. E. 1986. Plant drugs in the twenty first century. Economic Botany **40**: 279–288.

UNESCO. 1989. Resolution on the Safeguarding of Traditional Culture and Folklore. UNESCO, Paris.

United Nations. 1993. Commission on Human Rights. Sub-commission on prevention of discrimination and protection of minorities. Working Group on Indigenous Populations. 19–30 July 1993. First International Conference on the Cultural and Intellectual Property Rights of Indigenous Peoples. Whakatane, 12–18 June 1993, Aotearoa, New Zealand.

World Resources Institute. 1992. Global biodiversity strategy: Guidelines for action to save, study and use Earth's biotic wealth sustainably and equitably. World Resources Institute, Washington, D.C.

3

Collecting Ethnobotanical Data: An Introduction to Basic Concepts and Techniques

Miguel N. Alexiades
Institute of Economic Botany,
The New York Botanical Garden

Selected Guidelines for Ethnobotanical Research: A Field Manual, 53–94
Edited by Miguel N. Alexiades
© 1996 The New York Botanical Garden

Introduction

Much of the data collected by ethnobotanists is cultural data: when an informant describes the use of a plant, he or she is explicitly and implicitly referring to the concepts and categories of a particular individual with a particular cultural experience that may or may not correspond to the concepts and categories of the ethnobotanist. The term *naive realism* is used to describe the self-centered and almost universal belief that all humans define the world around them and categorize their experience in the same way (Spradley, 1979). When recording accounts on plant use, a worker not trained in cross-cultural analysis will unwittingly project his or her expectations, categories, and experience onto the informant's account. The seemingly straightforward method of collecting cultural data—talking to another person—belies the tremendous challenge of learning to recognize and minimize the ways in which we unconsciously reinterpret and reformulate the experience of others on the basis of our own. Here, I outline some basic anthropological concepts and field techniques relevant to the collection of culture-specific data in ethnobotany, as discussed in some of the literature (Bernard, 1988, 1994; Burgess, 1982a; Ellen, 1984; Pardinas, 1991; Spradley, 1979; Werner & Schoepfle, 1987). This overview does not attempt to present a thorough or systematic treatment of either the subject or the literature; rather, it hopes to introduce ethnobotanists with little or no background in the social sciences to a broad and critical topic (but see also Etkin, 1993). In addition, some pointers, based on my field experience and numerous discussions with colleagues, relating specifically to the collection of data on plant use knowledge are also provided.

Sources of Bias When Collecting Cultural Data

A scientist is not a blank slate capable of absolute objectivity, particularly when it comes to recording culturally based information. All individuals carry a certain amount of cultural "baggage" or "filters" in the form of preconceived notions, stereotypes, and expectations (Werner & Schoepfle, 1987). These biases are determined by a combination of the individual's culture, subculture, and life experience. Nationality, ethnic identity, socioeconomic class, ideology, age, gender, and profession are some of the variables that determine what we notice and how we interpret it.

Most of our concepts and beliefs, taken for granted as they are, are derived from our experience in our own society, and this experience is easily and inappropriately projected onto the acts, including spoken acts, of others. Anthropologists refer to "blind spots," "ethnocentrism," and "semantic accent" to describe the processes through which workers introduce their personal bias and distortions onto the interpretation of culture-specific behavior and knowledge (Werner & Schoepfle, 1987).

Another source of bias, reactivity, is introduced as the informant filters the information he or she provides in a conscious or unconscious reaction to the researcher's appearance and behavior (Bernard, 1988). An informant who suspects an interviewer of having hidden motives and consciously or unconsciously conceals certain types of information offers an extreme example of reactivity. Other forms of reactivity are more subtle. For example, a fieldworker might express particular interest in some forms of plant use and be uninterested in others. Over time, the conversations and exchange of information will tend to focus on those subjects that the researcher considers important, as informants will tend to elaborate on themes that are responded to with interest. Eventually, the representation of the ethnobotanical lore of the individual will be skewed by reactivity to the ethnobotanist's preconceived notions of what is important. Again, the combination of the context in which ethnobotanical data is collected, the social and cultural makeup of the informant and researcher, and the means that are used to eliminate subjective influences will all determine the degree and form of reactivity.

Finally, there is also a natural tendency to overlook that which is obvious or ordinary and to pay particular attention to the exotic or unusual. Inexperienced ethnobotanists in particular may fail to notice mundane but very important uses of plants, as construction materials, for instance, and focus instead on less ordinary uses, as hallucinogens, perhaps, which are also often harder to document in a meaningful way.

Fieldworkers can minimize these biases through the use of techniques specifically designed to collect cultural data. In addition, explicitly identifying the ways in which a study is biased helps researchers present their results in a meaningful perspective.

Interviewing

The interview, in its various forms, constitutes the basis of most ethnobotanical data collection. Insofar as an interview consists of two or more people talking to each other, the process has a simple and straightforward appearance. In reality, the way in which the interview is conducted, how questions are constructed and presented, and the answers recorded, all have significant impact on the quality, quantity, and meaning of the data collected. It is through the interview that the fieldworker can record the different epistemological, symbolic, and pragmatic aspects of plant use and situate this information in a meaningful context.

Interviewing Protocol

Interviewing is a dynamic process involving spoken interactions between two or more people. Its value as a tool of inquiry depends on the context in which it takes place and on the interviewer's skills. Local rules of conversation have to be discovered and adopted, including the use of visual cues such as eye contact, posture, and so on. Such rules can be learned only with time and careful observation. Ethnobotanists should develop sensitivity and respect when individuals show shyness or reticence on certain subjects; personal boundaries often change as people get to know each other. The art of interviewing depends on the delicate balance of curiosity and respect and on finding the thin line between showing interest and asking too many questions. The best information is obtained over extended periods of time, where mutual trust and understanding can develop, allowing the researcher to cross-check his or her observations repeatedly.

It is impossible to do good ethnobotany without having a good rapport with informants, although what constitutes good rapport is culturally defined in each society. The ethnobotanist needs to learn the local, culture-bound features that build rapport. Spradley (1979) suggested that despite the intercultural differences in how rapport is developed, the ideal relationship between a fieldworker and an informant usually proceeds through stages, beginning with apprehension and ending in full participation.

During the early stages of rapport building it is best to discuss familiar subjects or plants. Spradley (1979) suggested that during the early stages of the relationship, workers should explain clearly and repeatedly their interests and motivations. As the researcher and informants become acquainted over time, the latter will feel safer and more comfortable about discussing topics in greater detail and will broach new, more sensitive, ones.

Ethnobotanists should not expect all people to want to share their knowledge. Clearly, individual or group decisions not to participate in the study need to be respected. Good interviewers are aware of the pace at which people feel comfortable in establishing a relationship and sharing their knowledge. In general, it is better to err on the side of appearing naive, by asking only "obvious" questions at first. Though sharing information obtained from other informants may serve as a prompt to elicit information and help build rapport, its use should be particularly limited during the early stages of an informant–researcher relationship. Modesty about one's knowledge is usually most appropriate: interviewers should always demonstrate awareness of the fact that informants know more than they do.

Asking Questions

For a question to be valid it has to have the same meaning to the receiver as to the sender; that is, it has to be formulated in a culturally meaningful and appropriate way. Asking an Amazonian shaman during an ordinary interview whether he knows how to practice sorcery is usually as legitimate as asking an American in a street interview if he or she shoplifts. In this case, the question is not legitimate because it violates the rules for acceptable social intercourse. Questions are also illegitimate when they are based on concepts or assumptions alien to the system of meaning of the culture in question. Asking a tribal healer which plants can be used to reduce high cholesterol, for example, is probably asking an illegitimate question, because the category cholesterol most likely does not exist in his or her culture. Even if the term does exist in the culture, it probably has a different meaning. The problem is thus not only finding the correct idiom in which to express the question but also finding the correct subject or question to ask in the first place (Kemp & Ellen, 1984).

Failure to do so will result in data that simply reflect the mental contrivances of the ethnobotanist, since they will be based on answers rooted in a misunderstanding or in the desire to please or save face. Readers can consult Foddy (1993) for an additional discussion of the different theoretical and practical aspects related to asking questions in interview settings.

The use of native language to collect data is usually also a prerequisite to detailed analysis, as language is a major cultural filter: there are many concepts and categories that have no semantic equivalent and that may easily be inappropriately translated by an unskilled interpreter. Familiarity with the way informants think and speak is the key to successful communication, allowing one to pose meaningful questions and to interpret answers correctly. Asking—both constructing and phrasing—culturally legitimate and answerable questions is a particularly hard task for the beginning fieldworker, who typically lacks linguistic and cultural competence in the group under study.

In general, questions should not be too complicated, but then again they should not be too short or ambiguous (Kemp & Ellen, 1984). Interviewers need to be particularly wary of asking leading questions—questions that explicitly or tacitly suggest an answer. Leading questions convey to informants that there are specific expectations as to the content of their reply. This perception may cause them to provide distorted or erroneous information, again consciously or unconsciously, in order to please. Asking leading questions is the most common interviewing error, and it is harder to avoid than one might think. Examples of leading questions include, "Is this a medicinal plant too?" "When, at night?" and "What do you mean by *calentura,* fever?" Experienced interviewers can consciously and selectively use leading questions as prompts or checks, but doing so requires skill and a good understanding of the cultural norms in question. Bernard (1988) argued that all questions are to some extent leading questions and suggested that learning to interview is not so much about learning how to avoid asking leading questions as it is about learning how to ask them properly. One of the challenges for ethnobotanists, who often have a clear agenda of the kind of information they wish to obtain, is to create a space in the interviewing process where the informant will feel that it is okay not to have an answer or to provide a specific answer.

Interviewers should not interrupt or jump in with questions. During normal social intercourse, interruption occurs because one person becomes impatient and needs to express himself or herself. This behavior is not justified in a research interview (Whyte, 1982). Interruptions are not only rude, but they may also lead to loss of information. Sometimes it hard to refrain from interrupting, because each culture has its own pace of conversation. Urban North Americans and Europeans generally avoid prolonged periods of silence during most forms of social intercourse, but in other cultures lengthy pauses are the norm (Bernard, 1988). Thus, what may appear to the interviewer as silence after a reply (and an appropriate time to speak) may actually be a "pregnant pause," during which the informant is collecting some more thoughts on the matter. In fact, periods of silence after a reply can be used as a "silent probe" to elicit more information (Bernard, 1988). Interviewers should learn to cultivate patience and notice the pace of conversation around them. In general, ethnographers recommend waiting at least 15 seconds in silence before asking a new question. Clearly, this is only a generalization, and appropriate waiting intervals may be shorter or longer in different settings and among different cultures (Ebihara, pers. comm.). Interviewers may occasionally encounter an informant who appears to talk endlessly, diverging considerably from the subject. Deciding when a reply becomes "excessive" is a judgment call, and if interruption is necessary, it should be done tactfully (Whyte, 1982).

In general, informants should not be contradicted. Although contradiction is sometimes strategically used by skilled interviewers as a specific technique to elicit certain responses (Kemp & Ellen, 1984), it should be avoided by all except competent interviewers certain of what they are doing. This does not mean, however, that informants should not be asked to clarify a point or an apparent contradiction between statements, particularly if there is a good rapport between the informant and researcher.

Interviewers should not be disapproving or judgmental of informants or the information they are giving (watch for unconscious reactions such as facial expression). Showing or expressing doubt about certain beliefs may alienate the informant and create misgivings as to the interviewer's suitability as a recipient of the informant's knowledge. This reaction is particularly likely in rural areas, where people are accustomed to, and wary of, an all-too-common urban contempt for their culture. On the other hand, overenthusiastic reactions should also be avoided. There is a fine balance between showing interest as a way of building rapport and prompting more information and overreacting or expressing clear expectations as to what the informant should (and therefore should not) say.

Types of Questions

There are several types of questions that may be used during an interview. The ability to choose the right type of question at the right time is one of the key skills in successful interviewing. **Open questions** leave the informant with considerable "width" to answer; for example, "Tell me about this plant." Open questions are, in effect, a "catch-all," or residual category (Whyte, 1982). They are particularly useful at the beginning of an interview to "break the ice" and to identify topics of interest. Open questions exert very little control over the informant's responses and are thus an important instrument in ethnobotanical interviews. **Indirect questions** are a way of "beating around the bush to find out what is in the bush" (Werner & Schoepfle, 1987: 305). The nature of these questions is clarified by the context in which they are formulated. Indirect questions are often employed when dealing with sensitive subjects, such as death, sex, or sorcery. For example, asking a North American, "Was he sick for a long time before passing away?" may be used as an indirect way of probing into the cause of death.

   Direct questions are also known as the "wh" questions—what, when, who, whom, why, and how (Werner & Schoepfle, 1987). Although their phrasing is quite flexible, these questions exert considerably more control over the response than do open or indirect questions. **Closed questions** require a yes or no answer and are usually asked using fixed phrasing. They exert the most control over the response, so they should be employed very carefully. In general, the less one knows about an informant or his culture, the less one knows what the appropriate questions are or how to phrase them and the more "open" one's questions should be (Cohen, 1984).

Probing

The interviewer's manner of talking and interacting should encourage informants to speak freely and informatively. Prompting or probing can be used to direct informal and semistructured interviews or to rekindle a conversation. The various types of probes provide different degrees of directiveness or control (Bernard, 1988; Whyte, 1982). A simple nod or saying "uh-huh" often encourages the informant to speak more while exerting little influence on the direction of the conversation. Even a period of silence following a reply can serve as a probe, indicating to the informant a willingness to hear more on the subject.

Repeating the last phrase or sentence the informant said with a rising inflection also can serve as a probe. Probing by making a statement or raising a question on a previous remark made by the informant still follows the informant's lead but introduces more control than the previous probes. Referring to a remark made in the past by an informant implies exerting a greater degree of control than does referring to a remark that has just been made. Introducing a new topic is the most directive probe (Whyte, 1982).

Other prompts include tactical use of statements of disbelief or playing dumb—counterproductive in most contexts (Kemp & Ellen, 1984)—and the use of projective aids such as cards, pictures, photographs, or herbarium specimens. The "phased assertion" or "baiting" technique depends on acting as if you already know something in order to encourage people to open up (Bernard, 1988). Again, as with all probes, this needs to be employed selectively, ethically, and with tact.

Types of Interviews

The preceding sections provided some pointers on how to conduct interviews and ask questions. The setting of the interview and the degree to which questions are predetermined, directive or closed all exert different degrees of control on the interview (Kemp & Ellen, 1984). It is important to understand the ways in which control is exerted during an interview and to recognize how much control is appropriate given the interview's aim and context. The type of interview conducted, as well as the questions asked, affects the degree of control.

Informal Interview

Lacking any structure or control, the informal interview is at one end of the spectrum of interview types. The researcher simply makes notes, either during or after a casual conversation, of what is said or observed. Although informal interviews will usually take place during the whole study (whenever one talks to an informant), they are most useful at the early stages of a study while getting acquainted with an area.

Unstructured Interview

The unstructured interview has the appearance of a casual conversation, but both the informant and interviewer are aware it is an interview. The interview develops within a framework established by the researcher. The idea is to get the informant on a topic while exercising minimum control over the responses, that is, the actual content of discussion. The rule here is "get an informant onto a topic of interest (e.g., medicinal plants) and get out of the way" (Bernard, 1988). Thus the unstructured interview is flexible but controlled (Burgess, 1982b). The unstructured interview is particularly useful in building initial rapport with informants, approaching sensitive issues, conducting in-depth studies of cultural aspects of plant use, and developing formal guidelines for semistructured interviews or questionnaires (Bernard, 1988).

Semistructured Interview

The semistructed interview has much of the flexibility of the unstructured interview but is based on the use of an interview guide: a list of questions and topics that need to be covered, usually in a particular order (Bernard, 1988). The interviewer still is free to follow leads but has a specific agenda. This form of interview is particularly useful once specific research questions are identified and need to be pursued in greater detail. Many of the interviews relating to medicinal plants, in which mode of collection, preparation, and administration are discussed, can fall under the category of a semistructured interview.

Structured Interview

Structured interviews are based on a set of fixed questions that are presented to several informants, often in the form of a questionnaire (Bernard, 1988; Burgess, 1982b). A number of structured interviewing techniques used in cognitive anthropology are directly relevant to ethnobotany (Werner & Schoepfle, 1987). These include **free listing,** in which informants are asked, for example, to name as many plants as they can in a given period of time. **Triad tests** involve giving informants three things and telling them to "choose the one that doesn't fit," or "choose the two that seem to go together best." The "things" can refer to actual plants, photographs, herbarium sheets, or picture cards. Informants can also be asked to **rank** plants in order of importance within a certain context. For example, informants who provide a list of plants used to make house beams could be asked which one is "best" or used most often, and why.

Because structured interviews are based almost entirely on direct and closed questions, their weakness lies in their vulnerability to the introduction of interviewer bias through the use of inappropriate questions. On the other hand, data collected through structured interviews can be subjected to quantification and statistical analysis (see Phillips, Chapter 10, this volume). In the event that structured interviews are used, they should perhaps be reserved for the later stages of the ethnobotanical study, when the researchers are fairly confident in their understanding of the local culture and in their ability to pose meaningful questions.

Techniques of Inquiry

Techniques are tools, and the choice for using one over the other depends on the aims and theoretical approach of the study, field conditions, and expertise of the researcher. The following are a few of the many techniques for collecting data that are available to ethnobotanists. Each employs one or more of the interview types listed above.

Direct or Participant Observation

As the name implies, direct or participant observation is based on observing human–plant interactions, such as which plants are collected and used for specific purposes. The ethnobotanist accompanies the people with whom he or she is working, often directly participating during such tasks as gathering of fruits or forest products, hunting, farming, or collecting and using medicinal plants. During this time, the ethnobotanist may, if appropriate, record observations, ask questions, and collect voucher specimens.

Because the information collected through participant observation is highly reliable, this technique is particularly valuable for documenting plant use. In medicinal plant studies, the presence of a qualified medical practitioner could provide valuable complementary etic diagnoses of symptoms before and after treatment. The value of participant observation as a research tool depends greatly on the ability to notice details and nuances. It is very easy to ignore that which appears obvious or commonplace but which may be extremely important (Bernard, 1988; Holy, 1984).

Simulation

Simulation is a technique used to get participants to reenact activities that are no longer performed or to perform them out of context (Clammer, 1984). The technique is considered valid as long as participants are able to remember accurately what to do and provided it is socially or psychologically possible to perform the actions, given the "artificial" context. Ethnobotanical simulations could include observing the manufacture of outmoded plant-based artifacts or the preparation of a rarely used remedy.

Field Interview

Also referred to as a "bagging interview" (Alcorn, 1984), "walk-in-the-woods interview" (Phillips & Gentry, 1993a), or "ethnobotanical inventory" (Boom, 1987), the field interview consists of walking in one or more vegetation zones with an informant, collecting and taking notes on plants and their uses. The selection of plants for discussion may be decided upon by the informant, by the ethnobotanist, or by both, depending on the degree of control the ethnobotanist wishes to exert over the choice of plants and subjects discussed. In the most directive of field interviews, informants are taken to plots in one or more vegetation zones where all or certain plant individuals have been previously marked or numbered (see, e.g., Balée, 1986; Carneiro, 1978; Pinedo-Vásquez et al., 1990). Because all informants are shown the same individual plants, data collected in this fashion are amenable to certain types of quantitative analysis (see Phillips et al., 1994; Prance et al., 1987). The strong directiveness of the approach, on the other hand, increases the risks of reactivity and other sources of bias.

The advantages of field interviews include the facts that informants get to see the plants in their natural state, thus minimizing the risk of misidentification, and that the context of the interview itself can lead to the discovery and discussion of new important questions. The time-consuming nature of the procedure does place, however, a constraint on the number of informants and plant species that can be included in the sample.

Plant Interview

In a variation on the field interview, one or more plants can be collected in the field, brought back to the village, and presented to informants (Alcorn, 1984; Boom, 1987). Pressed plant specimens can also be used in this way. Because the method is less time-consuming than field interviews, the plant interview allows more informants to be included in a given period of time. Furthermore, it allows the possibility of working with some of the older members of the community, who may have difficulty walking to a given vegetation zone.

One disadvantage of the technique is that small sections of large shrubs and trees may not be readily identifiable by informants, who often rely on such morphological, architectural, or ecological characters as plant height, crown shape, branching pattern, presence of latex, habitat, and the association with other plants and animals for positive identification (Berlin, 1992). This problem is likely to be compounded when pressed material is used, because the spatial arrangement of organs, as well as the original colors and smells, are not readily retained. In some cases, it may be possible to complement a pressed specimen with good-quality color photographs of the plants. The approach may be particularly useful during preliminary or short studies as a way of reconfirming or gathering complementary data or in order to increase the number of informants.

Artifact Interview

In the artifact interview, several informants are shown, sequentially or simultaneously, one or more artifacts and asked which plants are employed in their manufacture or preparation. For example, interviewers can point to different parts of the house, recording their names together with those of the plants and parts used to make them. These plants can subsequently be collected using the same informants (Boom, 1987). This technique is often a good way of beginning an ethnobotanical study in a community, as it is simple and will familiarize the community with the researcher's interests and methodology.

Checklist Interview

Informants may also be presented with a checklist of plant names and asked for their uses. This may be an interesting alternative method for well-known plants, but it is open to introducing errors due to frequent variations in taxonomic systems used by different groups (see Bennett & Gomez, 1991). Once again, visual aids, such as photographs, drawings, or herbarium sheets, may be used as complementary aids during the interview. Another possibility is to ask which plants are used to fulfill specific needs (e.g., thatching materials, fuelwood, remedies for diarrhea), or broad needs (e.g., food, construction materials). The degree of control exerted by the interviewer in this type of interview increases the risk of bias due to reactivity and use of inappropriate categories. Thus, interviewers need to ensure that the categories used to elicit information are culturally meaningful; that is, they are locally used terms. This approach can be used as part of the structured interviewing techniques discussed above.

Group Interview

The ethnobotanist can choose between conducting interviews with one or with several informants at a time; both approaches have certain advantages and disadvantages. Working with a group of people can be a highly effective and stimulating exercise, since individuals may "prime" each other to provide information. Some people will be more willing to share and "open up" in a group environment, others less.

Group discussions can produce a wealth of data and lead to discovery of new topics and questions. They can also serve as social occasions to facilitate the transmission of cultural knowledge across generations, a desirable outcome in many communities undergoing rapid deculturation. At other times, however, individuals may be reluctant to disclose certain types of knowledge in front of others—persons of the opposite sex, or rival clans or families, or children, for example. In these cases, a group setting might actually inhibit communication. In some cases, this problem can partly be solved by conducting group interviews with informants separated by age, gender, or whatever variable appears to be limiting to verbal communication in a group context. In any event, an understanding of basic community dynamics is often helpful in determining how and when to set up group interviews.

Group interviews may not be compatible with certain research goals, such as studying distribution and variation of ethnobotanical knowledge within a mixed group of people or when applying statistical tests that require the recording of responses as "independent events" (see Phillips & Gentry, 1993a).

Market Survey

Many cities and town markets have sections where medicinal plants, fruits, and other plant products are sold. These markets are often rich sources of ethnobotanical information (Bye, 1986; Bye & Linares, 1983, 1987; Martin, 1992; Van den Berg, 1984). The accessibility of markets, the large number of people involved, and the public nature of the market space itself offer favorable conditions for fieldwork. Direct observations, interviews, and surveys of both merchants and buyers can be used to obtain a broad range of qualitative and quantitative data concerning cultural, social, and economic aspects of plant product selection, use, and commercialization.

Data Collection and Transcription
Audio and Video Recording

Audio cassette recorders are essential for recording indigenous plant names and are also a valuable aid during interviews. Taping is faster and more accurate than note taking and allows interviewers to maintain a more free-flowing conversation during the interview. This advantage might be especially useful during field interviews or while collecting plants, when workers need to keep their eyes on the path or on their hands while collecting a plant.

Even when using a recorder, the researcher should jot down notes as a backup, in case of unnoticed battery or mechanical failures. Tape recorders used in the field produce the clearest recordings when a small external microphone, clipped to the shirt or jacket, is used.

Taped interviews can be transcribed and used as backups for notes taken during the interview or for future reference or analysis. When deciding whether to tape, bear in mind that full transcriptions of interviews, while extremely useful in many cases, are extraordinarily time-consuming: at least 6 hours of transcription for every hour of tape (Bernard, 1988). Furthermore, the use of cassette recorders is not always appropriate, as many people object to being recorded or feel self-conscious. There might also be certain subjects that people feel reluctant to discuss while being recorded. Suspicion and uneasiness may diminish as people gain trust in the fieldworker and his or her motives. Recording should not be done without obtaining prior consent, not only because doing so would be unethical but also because being "discovered" taping without permission may justifiably fuel additional suspicions and mistrust.

Video recorders, especially Hi-8 camcorders, are becoming increasingly accessible and portable and may serve as valuable tools to record sequences of plant collection and use. Because video recorders are more conspicuous and less common than cassette recorders, they may raise more suspicions. On the other hand, video or audio recording equipment sometimes provokes the opposite reaction; people may see them as statements of affirmation of the value of their knowledge or culture. Indeed, video technology has recently been successfully manipulated by tribal societies in their interactions with the national and global society (Asch et al., 1991; Turner, 1991). People often are thrilled to see their images or hear their voices and songs. Audio and video recorders can become valuable projective and educational aids, serving to elicit responses from other informants or encouraging reassessment of local knowledge.

Photographs may be an additional valuable aid in ethnobotanical research. Furthermore, giving informants copies of photographs of themselves and their families is often a good way of establishing reciprocity. As with video and tape recorders, permission should always be sought before taking pictures.

Field Notes

When taking notes during interviews, record verbatim transcripts of responses whenever possible. Interviewers should be very careful to distinguish between observations and interpretations: that is, what people say or do versus what the interviewer thinks they mean by their actions or words. Observations describe an act, including a spoken act. Interpretations project meaning into the act. Since meaning is socially constructed, the same act can have different meanings in different cultural settings (Holy, 1984). Upon first arriving in an area, the researcher will tend to base interpretations on his or her system of reference, not on that of the group under study. As the researcher gains insight into the underlying network of rules, logic, and meaning of the culture under study, interpretations will become more reliable.

Although researchers cannot or should not stop interpreting observed acts, it is important that they distinguish between interpretations and observation, both in their minds and in their field notes. Interpretations, personal reactions, or comments can be recorded in the field notes using brackets or some other such means, in order to visually distinguish them from observations. Questions should also be transcribed as close to verbatim as possible, again distinguishing them from responses by using backslashes or some other standardized means.

Notes taken in the field usually consist of short, abbreviated jottings, which can later be used to jog the memory and elicit other information ("mental notes"). It is important to transcribe these "field joggings" or "scratch notes" (Sanjek, 1990) into proper field notes as soon as possible: the longer the interval between note taking and transcription, the less accurate the memory the notes will be able to jog. If possible, a small pocket notebook should always be kept handy to record spontaneous insights, events, or information. Although most people get used to seeing a researcher constantly jotting things down, there may be times that taking notes is not appropriate. In these cases, a mental note can be made and jotted down soon after the event, but especially before going to sleep, as there is a considerable loss of memory or confusion of events overnight. A final point concerning note taking: informants may disclose information confidentially with explicit or implicit instructions on the need to maintain secrecy. It is considered unethical to disclose this type of information (Society for Economic Botany, 1994: app. 1). The challenge of learning to recognize confidential information increases as personal relationships develop with informants. If you are not sure whether a communication is confidential, ask the informant for approval before recording or disclosing information that might be sensitive, secret, or compromising.

A key tool in ethnography with potential value in ethnobotanical studies is a journal with daily entries, in which general remarks, notes on progress made with the research, and any other significant matters are recorded (Werner & Schoepfle, 1987). The journal may serve as a valuable source of information at a later stage and can be used to cross-check with interview transcriptions for clues as to how different aspects of fieldwork influenced each other and where biases may have been introduced. Journals are used by ethnographers as a principal way of gaining awareness of personal subjectivity, bias, and semantic accent (Werner & Schoepfle, 1987).

Workers should make photocopies of their field notes and journal as regularly as possible, particularly before traveling. If photocopying facilities are unavailable, the use of carbon copy notebooks is highly recommended.

Portable computers have become an increasingly accessible option for field data entry. Powerful machines allowing for considerable data storage and processing in the field are widely available at a reasonable cost. In remote locations, they can be powered with the use of a 12–16 watt solar panel and a 12-volt rechargeable power pack, car or dry cell battery, coupled with an inverter to convert the battery DC current into AC current for the computer. Some manufacturers also sell DC adapters. The advent of communications technology also allows data to be relayed back and forth from a remote field site via modem.

Recording Indigenous Plant Names

Plant names contain a wealth of information on how a particular culture perceives and utilizes its plant resources and on how plants or their uses are diffused (Balée, 1989; Balée & Moore, 1991; Berlin, 1992; de Candolle, 1898; Conklin, 1954). Training in articulatory phonetics is necessary in order to detect and correctly transcribe all sounds present in human speech, of which any one language employs but a fraction. Poorly transcribed names are of limited value and, in some cases, may actually create confusion for other workers using the data or subsequently visiting the area.

Fieldworkers planning to work with an unfamiliar language should acquire basic linguistic skills and work with or seek the advice of a linguist familiar with that particular language (see Appendix 2). At the very least, indigenous plant names should be recorded in audio cassettes for subsequent transcription by a linguist familiar with the language in question. If recorders are also used to collect ethnobotanical information, one should not expect linguists to spend hours listening to these tapes to transcribe a few plant names. A more efficient way of recording plant names for transcription is to go over a list of plants with an informant, thereby obtaining a continuous, uninterrupted recording of plant names with no other information. One should ask informants to repeat each name slowly at least twice. Ideally, each plant name should be recorded using at least two informants, thus compensating for any individual lisps or variation in speech.

Recording Medicinal Plant Uses

Medicinal plants are used to treat illnesses and not just diseases. *Illness* refers to the personal and social experience of being ill; as such, it is socially constructed and, unlike disease, is not necessarily identifiable from a biomedical perspective (Eisenberg, 1977; Pelto & Pelto, 1990). For example, attempts to describe Latin American culture-bound syndromes such as *susto, saladera,* and *ataque de nervios* in medical terms have not been wholly successful (Crandon, 1983b; Simons & Hughes, 1985).

The facts that disease and illness categories do not necessarily overlap and that medicinal plants are an integral part of a medical system, itself a cultural system (Kleinman, 1980), suggest that attempts to document the use of medicinal plants should consider the social and cultural context in which this use is embedded (Elisabetsky, 1986; Etkin 1988, 1990; Herrick, 1983). A meaningful discussion of medicinal plant use will often require an understanding of how the medical system defines therapeutic efficacy, as well as of what categories and explanatory models are used to define a particular illness and its treatment. The fact that most indigenous medical systems draw strong links between the health and illness of the individual body with that of the social body (Fabrega, 1971; Lock & Scheper-Hughes, 1990) further stresses the need to consider the social context in studies of medicinal plant use. Finally, medical systems, like any other aspect of a culture, are not fixed entities but dynamic systems undergoing constant change as they adapt and respond to changes in their cultural, political, and ecological universe (Crandon, 1983a, 1986).

Medicinal plant use information collected by fieldworkers with no training in cross-cultural analysis is likely to be superficial and heavily distorted by personal bias. For this reason, an inter- or multidisciplinary approach is strongly recommended (Elisabetsky, 1986; Etkin, 1988, 1993). When use of a multidisciplinary approach is not possible, fieldworkers should attempt to gain some basic understanding of the cultural basis of illness (see Appendix 2) and consider their own limitations when gathering, interpreting, and presenting such information. The following are some general remarks to bear in mind when collecting ethnomedical information.

Ethnobotanists cannot assume a priori that there is a direct correspondence between local and biomedical disease categories, even if there is an overlap in the terms used in the two cases. Confusion often arises, for example, when informants use biomedical terms such as *cancer, tumor,* or *rheumatism* to describe ailments, or *liver, kidneys,* or *ovaries* to refer to organs. Clearly, these terms cannot be assumed a priori to correspond to biomedical categories. In parts of Peru, for example, *mal de los riñones* ("kidney ailment") can correspond to lower backache, gastrointestinal disorders, kidney stones, or other disorders of the urogenital system (Neptalí Cueva, pers. comm.).

Similarly, many cultures have illness categories that have no direct translation in Western terms. Conversely, diseases considered to be different from a biomedical perspective may be considered by another cultural group to be symptoms of a single ailment. For example, in southeastern Peru, the disease *mal de arco* can correspond to skin fungal infections or to an outbreak of *Varicella zoster* (Neptalí Cueva, pers. comm.). Cholera, typhoid, stroke, rheumatoid arthritis, and acute consumption may all be locally diagnosed at certain times as *daño,* or sorcery (Alexiades, pers. obs.).

Ideally, direct observations of folk diagnosis and treatment by qualified medical practitioners can help define correspondence between local and biomedical disease categories. In many cases, however, matching the categories will not be possible. Researchers should clearly mark folk categories in their notes by underlining them, enclosing them in quotations, or using some other such means. It is also very important to record the diagnostic symptoms of the ailment: How does the informant know the person has "cancer" or "inflammation of the liver"? Diagnostic symptoms may be organic or behavioral. Different people or groups may have a different understanding of what these terms imply, and unless these details are recorded, the information will be largely meaningless.

Informant Reliability and Sampling Considerations

Within anthropology, it is generally accepted that there is often a contradiction between statements on behavior and actual behavior (Bernard et al., 1985): that is, there are differences between what individuals say, think, and do (Briggs, 1986; Murphy, 1971). Foddy (1993) presents an illuminating discussion of the limitations of verbal data, illustrating many of the ways in which the characteristics of the interview and questions asked affect the validity and reliability of responses. This problem underscores the importance of complementing data based on informant reports relating to behavior with behavioral data and direct observation (see Zent, chapter 10, this volume). Factors that may influence the "validity" of informant statements include involuntary or voluntary errors; the desire to please; ulterior motives; the informant's emotional state, values, and attitudes; and idiosyncratic factors of the interview situation itself (Freeman et al., 1987; Kemp & Ellen, 1984; Whyte, 1982). One measure of validity of plant use information may be to indicate whether the use is directly observed (highest validity), reported to have been used by the informant in the past, or reported to have been used by a third person not interviewed. These data may be quantified in the form of an index, with values decreasing from 3 to 1, respectively (Elliot, n.d.). Whereas *validity* refers to the accuracy of the information provided by an informant, *reliability* refers to the consistency of the information on a particular subject, both within and between informants (Pelto & Pelto, 1978).

In addition to posing questions and interpreting answers in a culturally meaningful way, there are formal means for establishing the accuracy and consistency of informant statements. These include checking for implausibility of statements and cross-checking data with and between informants. Triangulation, whereby the same question is posed in different ways and at different times to several informants, is an effective method for cross-checking and validating information (Dean & Whyte, 1959; Denzin, 1970; Vidich & Bensman, 1954; Whyte, 1982). For this reason, detailed ethnobotanical analyses require multiple interviewing sessions with the same and different informants.

Romney et al. (1986) present a mathematical model to estimate the cultural competence or knowledge of individual informants. The model is based on the idea that the agreement between informants is a function of the extent to which each knows the culturally defined "truth." This concept of "culture as consensus" has formed the basis of a number of quantitative analyses of ethnobotanical data (see Phillips, Chapter 9, this volume; Trotter & Logan, 1986), and it implies that widely shared knowledge is more important than knowledge that is not widely shared. The notion of culture as consensus raises two important and related methodological problems in ethnobotanical studies: the degree to which ethnobotanical knowledge is shared within a community or ethnic group and the problem of adequate sampling.

The question of intracultural variation has been ignored in most ethnobotanical and ethnographic studies (but see Alcorn, 1984; Bennett & Gomez, 1991; Berlin, 1992; Hays, 1974; Padoch & de Jong, 1992; Phillips & Gentry, 1993b). Pelto and Pelto (1975) review the historical, epistemological, and methodological reasons why most studies dealing with cultural knowledge and behavior, particularly among peasant and indigenous societies, have minimized intracultural variation. Ethnographic and ethnobotanical studies have traditionally relied heavily on the use of "key" informants considered to be the most knowledgeable in a particular domain of cultural knowledge. Healers, for example, are often assumed to be the main repositories of the group's knowledge of medicinal plants. Although healers may indeed possess a great deal of specialist knowledge on medicinal plants, other groups in the community may have their own unique repertoire of knowledge related to their particular medical epistemologies and roles in health care. Women, for example, play a key and unique role in health care, and their ethnomedical and ethnobotanical knowledge may not necessarily overlap with those of healers (Daniela Peluso, pers. comm.).[1] In addition, there often are different medical specialists within the same ethnic group, each with a unique store of knowledge (Cárdenas, 1989; Posey, 1992).

Studies that have focused on intracultural variation indicate that some realms of knowledge are universally, or at least widely, shared and others have more exclusive domains (Gardner, 1976). Although some variation in cultural and ethnobotanical knowledge may be dialectical (Berlin et al., 1974), idiosyncratic, or the result of the elicitation process itself (Foster, 1979), it seems that intracultural variation is often patterned (Hays, 1974) after such sociocultural variables as age (Phillips & Gentry, 1993b), gender (Burton & Kirk, 1979), culture change (Sanjek, 1977), profession, kinship affiliation (Boster, 1986), and the cultural or economic importance of the plant resource in question (Balée & Moore, 1991).

Ultimately, the study of intracultural variation can be used to draw important inferences about the nature and workings of ethnobotanical knowledge and processes (Berlin, 1992; Hays, 1974). Ethnobotanical descriptions should indicate areas of knowledge in which there is variation and, ideally, provide some explanation. In addition, it is only through repeated cross-checking and replication with the same informants that workers can determine the degree of consistency for each informant and filter out any "background" noise due to situational factors.

Intracultural variation presents additional methodological challenges to the ethnobotanist. For one thing, researchers need to make decisions relating to the type and size of sample of their

[1] Incidentally, the lack of studies documenting the ethnobotanical knowledge and behavior of women in folk and indigenous societies underscores the need for studies in this domain (but see, for example, Browner, 1985; Kainer & Duryea, 1992).

informants. The first step in this process is defining the sample universe, i.e., what and who will be studied. Geographical, ecological, and social criteria may all be used to define the boundaries of the study and thus of the sampling universe. Whereas some studies may include all the people living within a region or habitat, others may be based on a community or a smaller subsection of the community. Gender, age, and social roles, for example, may also be used to define the sampling universe. Once the researcher has chosen the sampling universe, decisions need to be made on how best to choose a *representative* sample from this universe. This, in turn, implies choosing the sample type and size.

In a **strict random sample,** every individual in the population has an equal chance of being selected. A random sample should not be confused with a haphazard sample. For a sample to be random, selection criteria must be truly objective and involve no conscious or unconscious selection on the part of the researcher. Haphazardly selecting households in a community is not a valid random-sampling technique, as the interviewer will most likely be biased by some aspect, such as location or appearance, even if he or she is not aware of how this bias is effected. Using a random-number table or choosing the *n*th person from the sample universe are common ways of selecting informants for a random sample.

A **stratified random sample** involves subdividing the sampling universe beforehand into important subgroups such as men and women, different age groups, and so on. In some cases, samples are subsequently drawn from these subpopulations according to their relative importance within the overall population.

In the actual field, strict random or stratified-random sampling is impossible to achieve. For example, individuals may be unavailable or unwilling to serve as informants. Oftentimes, informants are necessarily chosen haphazardly, opportunistically, or on the basis of a gradually built network of connections.

The question of sample size is also rather complex, and necessitates the making of conscious compromises. In general, the sample should be as large as possible. What constitutes an "adequate" sample will vary according to total population size, the degree of accuracy needed, and the type of data analysis to be performed. Thus, larger populations will require *proportionally* smaller sample sizes. On the other hand, important or significant processes should be detectable even with relatively small samples. In the case of statistical samples, there are standardized protocols available to determine sample size. For most other cases, there are no clear guidelines as to what constitutes an adequate sample size, and many factors need to be taken into account. Patterning of knowledge is an important factor, as greater degrees of internal variation in ethnobotanical knowledge or behavior require larger samples of informants. Researchers also need to take into account how representative their informants are and their credibility, questions which have been addressed earlier. High levels of inter-informant reliability would indicate that the sample size is sufficient for a given ethnobotanical question (Pelto & Pelto, 1978).

Clearly, many of the above decisions are conditioned by factors relating to the scope, objectives, and characteristics of the study and the site. The use of statistics to analyze data and formally determine correlations between variables requires specific conditions relating to sample type and size. The question of time is important, as it places clear constraints on sample size and frequently also on sampling methods. Shorter periods in the field usually imply a greater dependence on opportunism and serendipity to locate informants. Sample size is also determined by the depth to which individual knowledge and behavior is to be explored. Clearly, depth takes place at the expense of breadth and vice versa. The relative proportion of "key" and "general" informants in a study should be determined by how study depth and breadth are compromised.

As with other aspects of methodology, there is no one single approach or solution to the question of sample size and distribution. Rather, the responsibility of the researcher is to minimize any inconsistencies between the means and ends of the study. Furthermore, researchers should state explicitly the characteristics of their sample and their sampling methodology—that is, how informants were chosen and how many were used—for each aspect of their study. Finally, researchers will frequently be able to employ different sampling strategies for different aspects of their studies. Thus, a community-wide census might use a random sample whereas in-depth follow up interviews might be conducted with a panel of key informants selected on the basis of the availability, suitability, credibility, and representativeness of various individuals. Additionally, readers are encouraged to consult one of the many available specialized treatments of the topics introduced here (see, e.g., Babbie, 1992; Bernard, 1994; Bryman & Cramer, 1994).

Guidelines for Collecting and Recording Ethnobotanical Information

The following list includes some aspects of ethnobotanical knowledge that fieldworkers might consider useful when collecting information on plant use. The points covered are not all-inclusive, but they do include important questions that can be pursued to different degrees, depending on the objectives and scope of the study. Furthermore, these points are presented as a general guideline of topics of interest to ethnobotanists, not as a list of questions to be presented to informants. That is, some of the general questions listed may not be appropriate in certain contexts. For example, while the recommendations encourage ethnobotanists to, whenever possible, record specific amounts used in preparation and administration of herbal remedies, interviewers should bear in mind that in some cases such information might be so context-specific as to make generalizations inappropriate. Similarly, the categories used to present some of the information—medicinal, food and industrial uses[2]—are ethnocentric and partly artificial (see Etkin, 1983, 1986) and may not correspond to how plant product use is classified locally.

Vernacular Name

The language of each plant name should be indicated, and whenever possible a literal translation should be provided as well: e.g., *xáxa síe* [Ese-exa, "fruit-pungent-smelling"]. For cultivated plants, the entire name for the cultivar or landrace should be in-

[2]Another "category" of use considered by many ethnobotanical studies, that of "religious" or "magical" uses, is not discussed specifically, as the separation between "medicinal" and "religious" plants is often not recognized locally (see Brown, 1985; but see Balée, 1994: 5).

cluded: e.g., *maíz blando morado,* as opposed to simply *maíz*. Even for wild plants, several varieties or forms are often recognized. Workers should be careful to distinguish between the name given to the whole plant and the name given to the plant's part, organ, or product. For example, among rural inhabitants in the southeastern Peruvian Amazon, the Spanish term *krisneja* is used to refer to the thatched leaves of *Geonoma deversa* (Poit.) Kunth, whereas the name of the whole plant is *palmiche* (Alexiades & Peluso, pers. obs.). The following lists include pointers for gathering data related to different aspects of plant resource collection and use.

Plant Resource Harvest

Identification

- How does the informant recognize the plant (e.g., morphological, ecological characters)?
- Are there different forms, varieties, or landraces? How are these recognized? Do they vary in their properties or characteristics (e.g., medicinal effect, taste, productivity, growing requirements)?

Parts Collected

- The parts collected are not always the same as the part(s) used. The folk term for the part collected and its botanical equivalent should be included.

Context of Collection

- Who collects (e.g., men, women, children)?
- When
    Do planting, collecting, and harvesting activities occur at certain times of the day, lunar cycle, or season?
    Is collecting or harvesting limited to a plant developmental stage?
    What are the planting and harvest dates for cultivated plants? How many days or months until maturity?
    What environmental cues are used to determine timing (e.g., "just after the first frost," "when the wild geese return")?

- How

  Are any rituals or special preparations performed before, during, or after collection or harvest (e.g., fasting, ritual purifications, incantations)?

  What tools or special implements are used for planting, collecting, or harvesting?

  Are fruits harvested by cutting down the fruit tree or climbing it, or are they collected from the ground?

- Where (e.g., are wild plants collected from particular environments only? See Ecological and Management Status: Distribution, below.)

Storage

- Is or can the collected plant or plant part be stored before use?
- How is it stored (fresh, buried, dry, etc.), and for how long?

Plant Resource Use: Medicinal[3]

Preparation and Storage

- Part(s) used
- Other ingredients (any admixes, solvents?)
- Amounts used. This will often be given in local measures (handful, liter, bottle, etc.). Measures may be converted at a later date, but conversion should be done while one still has the original measuring container.
- Processing method. Some commonly used processing methods are listed below along with Spanish equivalents. When several processing methods are combined, each should be included. The local terms for processing methods should be employed, and these may be followed by the English or standard equivalents, if indeed there is such a correspondence. Because the same terms sometimes have different meanings or may be meaningless in different areas or among different people even in the same area, workers should be sure to understand what the informant means by each term.

  Juice (*jugo, zumo;* often used interchangeably: *zumo* often re-

---

[3] Collectors may also consult Croom, 1983; Etkin, 1986, 1988; and Penso, 1980, for additional guidelines.

fers to the pure juice, and *jugo* usually refers to the juice diluted in some water). Plant part is crushed or squeezed and juice strained.

Crushed, mashed *(machucado)*

Powder *(polvo).* Plant part is usually dried (e.g., in the sun, oven, fireplace) and then pounded or ground, and possibly sifted.

Hot infusion, tea *(agua aromatica, infusión, mate, té, tisana).* Boiling water is poured over the plant part. The preparation is left to steep, usually for a few minutes but sometimes longer. It may be drunk hot or cold. Infusions are often used for leaves or other nonwoody tissues of medicinal plants.

Decoction *(decocción).* Plant part is boiled in water. Decoctions are often used for the woody parts (root, bark) of medicinal plants.

Syrup *(jarabe).* Plant part is boiled in water with sugar until a solution of the desired consistency is obtained.

Cold infusion, cold extract *(macerado en agua fría, extracto frío).* Plant part is steeped in cold water, often overnight, then the mixture is strained. Sometimes the plant part is crushed before or during the addition of water.

Alcohol extract *(macerado en alcohol, extracto de alcohol).* Plant part is steeped in alcohol or drinking spirits. A **tincture** consists of one part of the plant to several parts of alcohol left to steep for a few days in a tightly sealed container.

Ointment, cream, salve *(pomada, crema, unguento).* Processed plant part (e.g., crushed, powdered, infusion) is mixed with lard or fat.

- Storage. Is the preparation used fresh, or can it be stored? If stored, where (container, underground, etc.), how (dry, steeped in alcohol, etc.), and for how long?
- Variations. Note any variations depending on the specific circumstances, such as the nature of the patient or of the ailment.

Route of Administration or Application

- Internal
  Oral (drink or eat; see Preparation, above)
  Inhale vapor of decoction, infusion, or smoke

- External

  Chewing and spitting

  Steam or vapor baths

  Blowing or passing smoke over affected area

  Herbal baths

  Poultice *(cataplasma)*. Plant part (usually crushed or bruised and mixed with a little hot water) is applied directly over area, and area is covered with a cloth or rag. May be applied hot or cold.

  Plaster or poultice *(emplasto)*. Plant part (usually crushed or bruised and mixed with a little hot water) is sandwiched between two layers of cloth and the cloth is applied to the skin. May be applied hot or cold.

  Compress *(compresa)*. Plant part (usually crushed or bruised and mixed with a little hot water) is applied directly to the skin. May be applied hot or cold.

  Rubbing *(frotación)*. Plant part (usually crushed or mixed with water or processed as ointment or liniment) rubbed on body.

  Dosage or posology. Include amounts, frequency, and length of treatment.

Adjunct Therapies

Is treatment preceded, accompanied, or followed by any other treatment, therapy, or ritual? If it is, which, for how long, and why? Therapeutic treatments associated with medicinal plants include:

- Bathing. In river or with herbal baths
- Special diet, food or behavior taboos
- Massages
- Supplications, invocations, incantations, prayers, or recitations
- Healing rituals with a specialized practitioner. These may involve such practices as blowing tobacco, sucking the affected part, or the ritual use of hallucinogens.
- Other rituals (e.g., symbolic payment to spirit or saint)

Ethnomedical and Ethnopharmacological Aspects

- Disease name or term. Local category or term used to describe the disease and the literal translation.

- Disease etymology. Local explanation or cause for disease or condition.
- Symptoms treated. This information often helps elucidate the specific action of the plant. Sometimes the local term is synonymous with the symptoms treated (e.g., *dolor de cabeza,* "headache"). Personal observations of the medical condition of the patient, before and after treatment, are most valuable, particularly if provided by a trained medical practitioner. Once again, when asking informants about symptoms, interviewers should beware of interpreting technical or medical terms literally.
- Response to therapy. Informants may describe certain reactions to the therapy, such as high temperature, sweating, hallucinations, dizziness, nausea, or vomiting. Alternatively, there may be symptoms that disappear after treatment. Again, first-hand observations are most valuable. Does the informant have any information on adverse effects, contraindications (e.g., should not be used during pregnancy), or overdose information?
- Supposed pharmacological action(s). Unless corroborated by a medical practitioner and clinical tests, notes on pharmacological action are purely speculative. Although these terms may be used subsequently in the analysis of data, their use to record ethnobotanical knowledge may be misleading.

Plant Resource Use: Food and Industrial

Many of the points covered in the lists above for medicinal plants are equally applicable to food and industrial plants. Preparation and processing of food and industrial plants are often complex, in which case workers should include the details of the procedure. These may include mechanical (e.g., crushing, sifting), physical (e.g., charring, cooking, boiling, drying), or chemical (e.g., fermenting) processes. The account should also include the time each part of the process takes, any special implements used, and the local terms used for any processes and products involved. When recording information on industrial plants, workers should try to be as specific as possible: for example, the roof of a typical Amazonian rural house includes more than five named distinct types of beams.

Plant Resource Use Status

Information on the significance or value of a plant product or its use toward the well-being or economy of the community in question is extremely important, though not always easy to obtain. Phillips (Chapter 9, this volume) and Zent (Chapter 10, this volume) provide detailed reviews of techniques used to quantify the cultural significance and importance of plant products in a particular community or society. The following list suggests some types of information researchers may want to elicit to determine a plant's use status.

- Source of use knowledge. Where, how, or from whom did the informant learn the use?
- Percentage of the community involved in plant product collection and use
- Cost of use. Labor, time, and expenses involved in collecting resource
- Frequency of collection and use of resource. Is the plant product still used, or was it used only in the past? If it is still used, how often (year-round, at certain times only)? If it is not used, when was it last used, and what replaced the product?
- Alternative resources available and criteria for selection. Is the product prized or used only as a last resort? Why?
- Accessibility of the plant product (e.g., spatial and temporal distribution, ownership or use rights)
- Economic role. Is the plant product used for subsistence, as a source of cash, both?
- Economic importance. What is the contribution of the resource to individual or group nutrition, health, or income? How important is it in the diet (see Zent, chapter 10, this volume)? Is it consumed year-round or at certain times only? If at certain times only, when and why? How commonly are the materials used? If prized, why?

Plant Resource Ecological and
Management Status

Information on the ecology and management of the plant in question is extremely valuable in any ethnobotanical study. Peters (chapter 11, this volume) discusses the methods used to measure key ecological parameters of economically important plant resources. Although this manual is not intended to cover the methodology used to document and understand resource management strategies, a few key questions relating to this topic are provided.

Management Status

- Is the plant resource managed? If managed, how intensively (e.g., protected, cultivated, or domesticated)?[4]
- Is the cultivated plant native or introduced? If introduced, when, from where, by whom? What traditional varieties do introduced varieties replace, and why? Are there any traditional varieties that are no longer cultivated? Why? Are these cultivated anywhere else?
- Roughly what amounts of the plant are harvested or collected? Are there any perceptions as to the impact of harvesting on the wild population? Is the resource perceived to be more or less abundant than in the past? Has planting or collection of the plant product increased (or decreased) in past years, and if so, in response to what?

Distribution

- Wild plants. What is the plant's favored habitat? Is the plant widely distributed or rare? Does its distribution affect its accessibility, frequency of use, or degree of management?
- Cultivated plants. Where is the plant cultivated? Does it have specific growing requirements (soil, water, light)? In what kind of agricultural system (e.g., dooryard garden, new swidden, old swidden, monoculture, etc.) is the plant grown? Is it grown in association with other plants? If so, which? Why?

[4] Recent publications in tropical resource management highlight the fact that categories such as "wild" are to some extent artificial and belie the inherent complexity and fluidity of tropical resource management systems (e.g., Posey, 1985; Posey & Balée, 1989; Redford & Padoch, 1992). Not cutting down certain plants or otherwise protecting them, helping them grow by clearing surrounding vegetation or dispersing seeds are all subtle ways of "managing" plant populations. Ideally, the management status of the plant should be described by using the native terms.

Reproductive Biology and Propagation

- Can the plant be or is it propagated from stakes or cultivated from seed? Has it been transplanted or grown from seed in the past?
- What are the mechanisms of pollination and dispersal of wild plants?
- In the case of fruit plants, is the plant monoecious (male and female flowers borne on same individual) or dioecious (male and female flowers borne on different individuals)? Dioeciousness can have a significant impact on the distribution, abundance, and hence value of a fruit resource, since only female flower-bearing individuals will be used directly.

Phenology

- Does the fruit plant, for example, produce both flowers and fruits in certain seasons? Every year? Every other year? If possible, try to distinguish between individual and population-level responses. Individual trees of some species, for example, fruit biannually, but in a given population there will be some trees fruiting each year (Oliver Phillips, pers. comm.).
- Are flowering plants annuals, biennials, or perennials? When do seedlings emerge? When does flowering occur?

Plant–Animal Associations

- Do animals use the plant as shelter or for food?
- Are there mythological or symbolic associations of animals and plants (see Reichel-Dolmatoff, 1989)? These often point to important ecological relationships (Posey, 1983).

The Informant

Background information on the informant is an extremely useful complement to ethnobotanical data: it places the latter in the context of the individual's personal history and experience, adding considerable depth to the data and their subsequent analysis. Personal information must obviously be gathered with tact, and only if appropriate. Often, this information is collected gradually during the fieldwork.

Personal Profile

- Gender
- Age (approximate)
- History (e.g., where born, where raised, how long living in this community, etc.)

Sociocultural Profile

- Ethnic group, community. Include location.
- "Role" or status in the community. Is the individual known to participate in some activities more than others (e.g., hunting, gathering, farming)? Is the person an established specialist (e.g., healer)? What degree of involvement does she or he have with the market economy, commercial activities, or other groups? What is the economic level of the person relative to the community? What is the person's degree of prestige?
- If a healer:
  Type of practice (e.g., herbalist, bonesetter, midwife)
  Characteristics of practice. Is it a full-time occupation? Is payment received? Who is usually treated (e.g., family, community, surrounding community)?
  History of practice. For how long has she or he practiced? From whom was practice learned? Where?
  Method of diagnosis (e.g., diviner, dreamer, using hallucinogens)

Acknowledgments

I am grateful to Conrad Gorinsky and Didier Lacaze for introducing me to ethnobotany and for countless insights shared over the years. William Balée, Michael Balick, Bradley Bennett, Daniela Peluso, Oliver Phillips, Christine Padoch, Charles Peters, and two anonymous readers provided valuable suggestions while reviewing earlier drafts of this paper. David Williams provided many helpful suggestions for the collection of information on food and cultivated plants. Libbet Crandon and May Ebihara kindly assisted in the understanding and presentation of many of the anthropological concepts and techniques that are discussed.

Robin Goodman assisted in the compilation of literature. I alone, however, am responsible for any errors, misinterpretations, or omissions. The Institute of Economic Botany of The New York Botanical Garden and the National Cancer Institute supported the preparation of this paper.

Literature Cited

Alcorn, J. B. 1984. Huastec Mayan ethnobotany. University of Texas Press, Austin.

Asch, T., J. I. Cardozo, H. Caballero & J. Bortoli. 1991. The storey we now want to hear is not ours to tell: Relinquishing control over representation: Toward sharing visual communication skills with the Yanomami. Visual Anthropology Review **7(2):** 102–106.

Babbie, E. R. 1992. The practice of social research. Wadsworth Publishing, Belmont, California.

Balée, W. 1986. A etnobotânica quantitativa dos índios Tembé (Rio Gurupí, Pará). Boletim do Museu Paraense Emilio Goeldi **3(1):** 29–50.

————. 1989. Nomenclatural patterns in Ka'apor ethnobotany. Journal of Ethnobiology **9(1):** 1–30.

————. 1994. Footprints of the forest: Ka'apor Ethnobotany—The historical ecology of plant utilization by an Amazonian people. Columbia University Press, New York.

———— **& D. Moore.** 1991. Similarity and variation in plant names in five Tupi-Guarani languages (eastern Amazonia). Bulletin Florida Museum Natural History **35:** 210–262.

Bennett, B. B. & P. Gomez A. 1991. Variación de los nombres vulgares y los usos que dan a las plantas los indígenas Shuar del Ecuador. Pages 129–137 *in* M. Rios & H. Pedersen, eds., Las plantas y el hombre. Ediciones Abya-Yala, Quito.

Berlin, B. 1992. Ethnobiological classification: Principles of categorization of plants and animals in traditional societies. Princeton University Press, Princeton, N.J.

————, **D. E. Breedlove & P. H. Raven.** 1974. Principles of Tzeltal plant classification: An introduction to the botanical ethnography of a Mayan-speaking people of highland Chiapas. Academic Press, New York.

Bernard, H. R. 1988. Research methods in cultural anthropology. Sage, Newbury Park, Calif.

————. 1994. Research methods in anthropology. Qualitative and quantitative approaches. Altamira Press, Walnut Creek, California.

————, **P. D. Killworth, D. Kronenfeld & L. Sailer.** 1985. The problem of informant accuracy: The validity of retrospective data. Annual Review of Anthropology **13:** 495–517.

Boom, B. 1987. Ethnobotany of the Chacobo Indians. Advances in Economic Botany **4.** The New York Botanical Garden, Bronx.

Boster, J. S. 1986. Exchange of varieties and information between Aguaruna manioc cultivators. American Anthropologist **88:** 428–436.

Briggs, C. L. 1986. Learning to ask: A sociolinguistic appraisal of the role of the interview in social science research. Cambridge University Press, New York.

Brown, M. F. 1985. Tsewa's gift: Magic and meaning in an Amazonian society. Smithsonian Institution Press, Washington, D.C.

Browner, C. H. 1985. Criteria for selecting herbal remedies. Ethnology **24(1):** 13–32.

Bryman, A. & D. Cramer. 1994. Quantitative data analysis for social scientists. Rev. ed. Routledge, New York.

Burgess, R. G. (ed.). 1982a. Field research: A sourcebook and field manual. George Allen & Unwin, London.

————. 1982b. The unstructured interview as conversation. Pages 107–110 *in* R. G. Burgess, ed., Field research: A sourcebook and field manual. George Allen & Unwin, London.

Burton, M. & L. Kirk. 1979. Sex differences in Maasai cognition of personality and social identity. American Anthropologist **81:** 841–873.

Bye, R. A. 1986. Medicinal plants of the Sierra Madre: Comparative study of Tarahumara and Mexican market plants. Economic Botany **40:** 103–124.

———— **& E. Linares.** 1983. The role of plants found in the Mexican markets and their importance in ethnobotanical studies. Journal of Ethnobiology **3(1):** 1–13.

———— & ————. 1987. Usos pasados y presentes de algunas plantas medicinales encontradas en los mercados mexicanos. América Indígena **47(2):** 199–230.

Candolle, A. L. P. P. de. 1898. Origin of cultivated plants. D. Appleton, New York.

Cárdenas T., C. 1989. Los Unaya y su Mundo. Instituto Indigenista Peruano (IIP), Centro Amazónico de Antropología y Aplicación Práctica (CAAAP), Lima.

Carneiro, R. L. 1978. The knowledge and use of rain forest trees by the Kuikuru Indians of central Brazil. Pages 201–216 *in* R. I. Ford, ed., The nature and status of ethnobotany. Anthropological Papers No. 67. Museum of Anthropology, University of Michigan, Ann Arbor.

Clammer, J. 1984. Approaches to ethnographic research. Pages 63–85 *in* R. F. Ellen, ed., Ethnographic research: A guide to general conduct. Academic Press, New York.

Cohen, A. P. 1984. Producing data: Informants. Pages 223–229 *in* R. F. Ellen, ed., Ethnographic research: A guide to general conduct. Academic Press, New York.

Conklin, H. C. 1954. The relation of Hanunóo culture to the plant world. Dissertation. Yale University, New Haven, Conn.

Crandon, L. 1983a. Between shamans, doctors and demons: Illness, curing and cultural identity midst culture change. Pages 69–84 *in* J. Morgan, ed., Third World medicine and social change. University Press of America, Lanham, Maryland.

————. 1983b. Why *Susto?* Ethnology **22:** 153–167.

————. 1986. Medical dialogue and the political economy of medical pluralism: A case from rural highland Bolivia. American Ethnologist **13:** 463–476.

Croom, E. M. 1983. Documenting and evaluating herbal remedies. Economic Botany **37:** 13–27.

Dean, J. P. & W. F. Whyte. 1959. How do you know if the informant is telling the truth? Human Organization **17:** 34–38.

Denzin, N. K. (ed.). 1970. Sociological methods: A sourcebook. Butterworths, London.

Eisenberg, L. 1977. Disease and illness: Distinctions between professional and popular ideas of sickness. Culture, Medicine and Psychiatry **1:** 9–23.

Elisabetsky, E. 1986. New directions in ethnopharmacology. Journal of Ethnobiology **6(1):** 121–128.

Ellen, R. F. (ed.). 1984. Ethnographic research: A guide to general conduct. Academic Press, New York.

Elliot, S. n.d. Bioresources database of ethnobiology: Guidance for contributors. Unpublished manuscript. Bioresources Ltd., London.

Etkin, N. L. 1983. Malaria, medicine, and meals: Plant use among the Hausa and its impact on disease. Pages 231–259 *in* L. Romanucci-Ross, D. E. Moerman & L. R. Tancredi, eds., The anthropology of medicine: From culture to method. J. F. Bergin, South Hadley, Mass.

————. 1986. Multidisciplinary perspectives in the interpretation of plants used in indigenous medicine and diet. Pages 2–29 *in* N. L. Etkin, ed., Plants in indigenous medicine and diet. Redgrave, Bedford Hills, N.Y.

————. 1988. Ethnopharmacology: Biobehavioral approaches in the anthropological study of indigenous medicines. Annual Review of Anthropology **17:** 23–42.

————. 1990. Ethnopharmacology: Biological and behavioral perspectives in the study of indigenous medicines. Pages 149–158 *in* T. M. Johnson & C. F. Sargent, eds., Medical anthropology: A handbook of theory and method. Greenwood Press, New York.

————. 1993. Anthropological methods in ethnopharmacology. Journal of Ethnopharmacology **38:** 93–104.

Fabrega, H. 1971. The study of medical problems in preliterate settings. Yale Journal of Biology and Medicine **43:** 385–407.

Foddy, W. 1993. Constructing questions for interviews and questionnaires. Cambridge University Press, New York.

Foster, G. M. 1979. Methological problems in the study of intracultural variation: The hot/cold dichotomy in Tzintzuntzan. Human Organization **38:** 179–183.

Freeman, L. C., A. K. Romney & S. C. Freeman. 1987. Cognitive structure and informant accuracy. American Anthropologist **89:** 310–325.

Gardner, P. M. 1976. Birds, words, and a requiem for the omniscient informant. American Ethnology **3:** 446–468.

Hays, T. E. 1974. Mauna: Explorations in Ndumba ethnobotany. Dissertation. University of Washington, Seattle.

Herrick, J. W. 1983. The symbolic roots of three potent Iroquois medicinal plants. Pages 134–155 *in* L. Romanucci-Ross, D. E. Moerman & L. R. Tancredi, eds., The anthropology of medicine: From culture to method. J. F. Bergin, South Hadley, Mass.

Holy, L. 1984. Theory, methodology and the research process. Pages 13–34 *in* R. F. Ellen, ed., Ethnographic research: A guide to general conduct. Academic Press, New York.

Kainer, K. A. & M. L. Duryea. 1992. Tapping women's knowledge: Plant resource use in extractive reserves, Acre, Brazil. Economic Botany **46(4):** 408–425.

Kemp, J. H. & R. F. Ellen. 1984. Producing data: Informal interviewing. Pages 229–236 *in* R. F. Ellen, ed., Ethnographic research: A guide to general conduct. Academic Press, New York.

Kleinman, A. M. 1980. Patients and healers in the context of culture. University of California Press, Berkeley.

Lock, M. & N. Scheper-Hughes. 1990. A critical-interpretive approach in medical anthropology: Rituals and routines of discipline and dissent. Pages 47–72 *in* T. M. Johnson & C. F. Sargent, eds., Medical anthropology: A handbook of theory and method. Greenwood Press, New York.

Martin, G. J. 1992. Searching for plants in peasant marketplaces. Pages 212–223 *in* M. Plotkin & L. Famolare, eds., Sustainable harvest and marketing of rainforest products. Island Press, Washington, D.C.

Murphy, R. 1971. The dialects of social life: Alarms and excursions in anthropological theory. Basic Books, New York.

Padoch, C. & W. de Jong. 1992. Diversity, variation, and change in Ribereño agriculture. Pages 158–174 *in* K. Redford & C. Padoch, eds., Conservation of neotropical forests: Working from traditional resource use. Columbia University Press, New York.

Pardinas, F. 1991. Metodología y técnicas de investigación en ciencias sociales. Nueva edición corregida y aumentada. Siglo Veintiuno, México D. F.

Pelto, P. J. & G. H. Pelto. 1975. Intra-cultural diversity: Some theoretical issues. American Ethnologist **2:** 1–18.

———— & ————. 1978. Anthropological research. The structure of inquiry. Cambridge University Press, New York.

———— & ————. 1990. Field methods in medical anthropology. Pages 269–297 *in* T. M. Johnson & C. F. Sargent, eds., Medical anthropology: A handbook of theory and method. Greenwood Press, New York.

Penso, G. 1980. The role of WHO in the selection and characterization of medicinal plants. Journal of Ethnopharmacology **2:** 183–188.

Phillips, O. & A. H. Gentry. 1993a. The useful woody plants of Tambopata, Peru: I. Statistical hypotheses tests with a new quantitative technique. Economic Botany **47:** 15–32.

———— & ————. 1993b. The useful woody plants of Tambopata, Peru: II. Additional hypothesis testing in quantitative ethnobotany. Economic Botany **47:** 33–43.

————, A. H. Gentry, C. Reynel, P. Wilkin & C. Gálvez-Durand B. 1994. Quantitative ethnobotany and Amazonian conservation. Conservation Biology **8:** 225–248.

Pinedo-Vásquez, M. D. Zarin, P. Jipp, & J. Chota-Inuma. 1990. Use-values of tree species in a communal forest reserve in northeast Peru. Conservation Biology **4:** 405–416.

Posey, D. A. 1983. Indigenous knowledge and development: An ideological bridge to the future. Ciencia e Cultura **35(7):** 877–894.

————. 1985. Native and indigenous guidelines for new Amazonian development strategies: Understanding biodiversity through ethnoecology. Pages 156–181 in J. Hemming, ed., Change in the Amazon Basin: Man's impact on forests and rivers. Manchester University Press, Manchester, England.

————. 1992. Interpreting and applying the "reality" of Indigenous Concepts: What is necessary to learn from the natives? Pages 21–34 in K. H. Redford & C. Padoch, eds., Conservation of neotropical forests: Working from traditional resource use. Columbia University Press, New York.

———— **& W. Balée (eds.).** 1989. Resource management in Amazonia: Indigenous and folk strategies. Advances in Economic Botany **7.** The New York Botanical Garden, Bronx.

Prance, G. T., W. Balée, B. M. Boom & R. L. Carneiro. 1987. Quantitative ethnobotany and the case for conservation in Amazonia. Conservation Biology **1:** 296–310.

Redford, K. H. & C. Padoch (eds.). 1992. Conservation of neotropical forests: Working from traditional resource use. Columbia University Press, New York.

Reichel-Dolmatoff, G. 1989. Biological and social aspects of the Yuruparí complex of the Colombian Vaupés Territory. Journal of Latin American Lore **15(1):** 95–135.

Romney, A. K., S. C. Weller & W. H. Batchelder. 1986. Culture as consensus: A theory of culture and informant accuracy. American Anthropologist **88:** 313–338.

Sanjek, R. 1977. Cognitive maps of the ethnic domain in urban Ghana: Reflections on variability and change. American Ethnologist **4:** 603–622.

————. 1990. A vocabulary for fieldnotes. Pages 92–121 in R. Sanjek, ed., Fieldnotes: The makings of anthropology. Cornell University Press, Ithaca, N.Y.

Simons, R. C. & C. C. Hughes (eds.). 1985. The culture-bound syndromes: Folk illnesses of anthropological and psychiatric interest. D. Reidel, Boston.

Society for Economic Botany. 1994. Guidelines of professional ethics of the Society for Economic Botany. Society for Economic Botany Newsletter **7 (Spring 1994):** 10.

Spradley, J. P. 1979. The ethnographic interview. Holt, Rinehart and Winston, New York.

Trotter, R. T., II, & M. H. Logan. 1986. Informant consensus: A new approach for identifying potentially effective medicinal plants. Pages 91–109 in N. L. Etkin, ed., Plants in indigenous medicine and diet. Redgrave, Bedford Hills, N.Y.

Turner, T. 1991. The social dynamics of video media in an indigenous society: The cultural making and personal politics of video-making in Kayapo communities. Visual Anthropology Review **7(2):** 68–76.

Van den Berg, M. E. 1984. Ver-o-peso: The ethnobotany of an Amazonian market. Advances in Economic Botany **1:** 140–149.

Vidich, A. J. & J. Bensman. 1954. The validity of field data. Human Organization **13:** 20–27.

Werner, O. & G. M. Schoepfle. 1987. Systematic fieldwork. Ethnographic analysis and data management. Vol. 1: Foundations of ethnography and interviewing. Sage, Newbury Park, Calif.

Whyte, W. F. 1982. Interviewing in field research. Pages 111–122 *in* R. G. Burgess, ed., Field research: A sourcebook and field manual. George Allen & Unwin, London.

II
Collecting Plant Specimens

Introduction

Collecting plant specimens is a key component of any ethnobotanical research project. It provides both a frame of reference for understanding and analyzing indigenous knowledge and plants and a basis for comparative analysis.

Chapters 4 through 7 deal with techniques used to make good-quality herbarium specimens, a vital step in obtaining reliable scientific determinations. Chapter 4 reviews the general principles and equipment needed to collect most vascular plants. Chapters 5, 6, and 7 are related specifically to the collection of plants for which specialized techniques are needed. Palms, which are very important economically, can be difficult to collect because their leaves and reproductive organs often reach unusually large dimensions. Mushrooms and, to a lesser degree, bryophytes are also economically important groups that require special, though not complicated, measures if the specimens are to be adequately identified. Finally, chapter 8 presents guidelines for collection of bulk plant material to be used for phytochemical analysis.

4

Standard Techniques for Collecting and Preparing Herbarium Specimens

Miguel N. Alexiades
Institute of Economic Botany,
The New York Botanical Garden

Selected Guidelines for Ethnobotanical Research: A Field Manual, 99–126
Edited by Miguel N. Alexiades
© 1996 The New York Botanical Garden

Introduction

Properly collected voucher specimens of ethnobotanically important plants are essential for obtaining taxonomic identifications. In addition, voucher specimens provide a permanent record of information that can be reviewed or reassessed. Ethnobotanical information without adequately vouchered specimens has little scientific value, since vernacular names vary widely among individuals, ethnic groups, and geographical areas. Taxonomic determinations thus provide an important basis for systematizing ethnobotanical knowledge and serve as a critical link between folk and Western knowledge systems (Bye, 1986). This chapter summarizes standard techniques used to make herbarium specimens. For other discussions, readers may consult Archer, 1945; De Wolf, 1968; Fidalgo & Ramos, 1989; Fosberg, 1939; Fosberg & Sachet, 1965; Johnston, 1939; Ketchledge, 1970?; Liesner, 1985; Mori et al., 1989; Savile, 1962; Smith, 1971; and Womersley, 1981. General discussions on managing herbarium collections and herbaria can be found in Forman & Bridson, 1989; Lot & Chiang, 1986; and Mori et al., 1989.

The Herbarium Specimen

Herbarium specimens are the main tools for taxonomic identification. In most cases, determinations will be only as good as the specimens on which they are based. Collecting sterile (lacking flowers or fruits), insufficiently annotated, or poor-quality specimens is a waste of time and space and, at best, will yield incomplete or unreliable determinations. Indeed, many taxonomists refuse to identify poorly made or sterile specimens. Well-pressed specimens on the other hand, provide much scientific information and are valuable additions to a specialist's collection.

A good herbarium specimen consists of a dried, pressed section of a plant containing well-preserved vegetative and reproductive (flowers, fruits) structures. Plant specimens are mounted
for permanent storage on sheets of standard ragbond paper measuring about $11^1/_2 \times 16^1/_2$ in. (28.7×41.7 cm). In the bottom
right corner of the sheet there is a label containing information
on the plant, a description of its appearance, and the area where
it was collected. A small paper pouch attached to the herbarium
sheet is used to keep small pieces of the specimen that might
become dislodged with time, as well as extra flowers or fruits
purposely collected (Figure 1). Herbarium sheets may also include photographs of the live plant or its parts, maps of the collection area, and hand-written annotations by taxonomists.

Collecting Specimens

Tools and Equipment

PRUNING SHEARS (CLIPPERS)   These should be sturdy, of high-quality steel.
Pruners are often carried in a sheath attached to a belt for safety
and convenience.

BUSH KNIFE OR MACHETE   A bush knife, known as *machete* in Latin America
and *panga* in parts of Africa, is very useful for cutting trails
through thick undergrowth and for cutting large branches, fruits,
and so on. Longer blades allow delivery of more powerful
strokes, but they also are harder to control and so are considerably more dangerous than shorter blades; a blade about 16 in.
long is adequate for most purposes. Bush knives should always
be used with extreme caution; they can easily inflict very deep
cuts, especially if they are well sharpened. When cutting through
vegetation, always swing the knife in arcs leading away from the
body. Exercise caution, particularly when cutting through vines,
because the blade can ricochet off a stem. When not in use, bush
knives should be kept in a sheath.

FIELD PICK OR TROWEL   Particularly useful for digging roots or removing
the entire plant.

EXTENDABLE TREE PRUNERS   These consist of a number of telescoping or
attaching poles (wood, fiberglass, or aluminum) with a pair of

Paper pouch

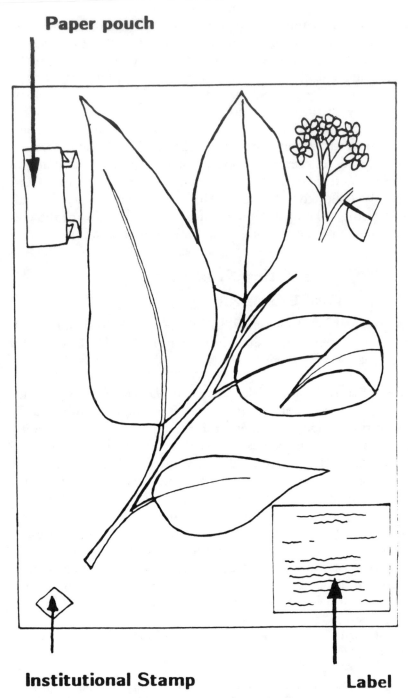

Institutional Stamp   **Label**

Figure 1. The herbarium specimen.

shears or a saw at the end. Telescoping poles can reach a maximum length of 12 m and are used for reaching the tree canopy. *Aluminum poles should not be used where there are any electric wires!*

COLLECTION NOTEBOOK This should be sturdy, of good quality, preferably made of weatherproof paper, and with a waterproof slip-on cover. Weakly bound books or spiral bindings, which may allow pages to slip out or tear easily, should not be used. Some collectors keep their collection notebook safe at the base camp and use a pocket notebook to jot down notes while collecting plants.

FIELD PRESS In most cases it is best to arrange the plant specimens in newspaper in the field, carrying them in a field press. A field press usually consists of two 12 × 18 in. plywood or wood lattice frames, held together with a pair of cords or web straps 4 ft long (Figure 2). Some of the more portable models look like a satchel made of lightweight waterproof fabric.

Specimens are folded in old newspapers or unprinted newspaper stock and slipped between sheets of blotting paper or thin corrugated cardboard or aluminum. To save space, one can place several

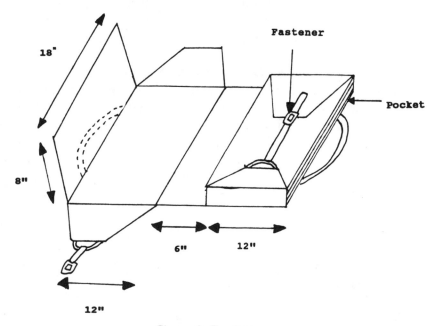

Figure 2. The field press.

newspapers with their enclosed plants between the corrugated sheets; a pair of rubber strands or cords hold the press together. Field presses are particularly useful when collecting few specimens, for transporting plants that wilt easily, and for small plants or fragile material that could otherwise be damaged or lost.

POCKET PRESS This consists of two pieces of cardboard, or an old notebook small enough to fit in a pocket, containing small sheets of blotting paper and held together by a strong elastic band. Pocket presses are used to collect delicate flowers.

NEWSPAPERS These are used for carrying specimens in the field press. Old newspapers or unprinted newsprint may be used, as long as they measure 11 × 16 in. or less; larger specimens will not fit in the 12 × 16 in. mounting sheet. Some tabloids are just the right size. Larger newspaper sheets may also be cut or folded to size.

LARGE, HEAVY-DUTY PLASTIC BAGS These may be used to carry the plant specimens to the base camp, where they can subsequently be pressed. Carrying plants to the base camp may be a good option when collecting time in the field is limited, when large amounts of material are collected, under extremely wet conditions, or when the plant material is sturdy enough.

SEALABLE PLASTIC, COTTON, AND BROWN PAPER BAGS (ASSORTED SIZES) Cotton or paper bags may be used to carry loose fruits and other fragile segments from the field to base camp. In some cases, the fruits can be dried in the same bags. Nylon net and cotton bags are ideal for storing fruits and plant parts after they have been dried. Plastic and, to a lesser extent, paper favor growth of mold. Sealable plastic bags can be used for transporting plant parts from the collection site to the base camp, as well as for a wide range of other purposes.

ALCOHOL Alcohol (70% solution or higher) is used to "pickle" fruits or flowers and to preserve bundles of plant specimens before drying. In many tropical countries, sugarcane alcohol can be obtained locally and cheaply.

PLASTIC WATERPROOF JARS These can be used to "pickle" fragile or rare flowers and fruits. An assortment of sizes should be carried. Glass jars are too heavy and fragile to transport.

PAPER TAGS These are used to label nonpressed, loose plant parts in the field, as well as whole specimens collected in plastic bags. The most convenient are 1-in. tags with short cotton strings.

PENCIL OR WATERPROOF PEN A soft pencil or wax crayon can be used to label news sheets. Field notes should be written with a no. 2 pencil or with a pen with indelible ink. Collectors planning to use a pen to label news sheets that will be soaked in alcohol should ensure that the ink does not run.

CAMERA Photographs can provide the specialist who will identify the plant specimens with valuable information on the plant's habit and the appearance of its parts. Furthermore, the photographs can be glued to the herbarium sheet, providing valuable information for future reference. Ideally, large leaves, such as those from some palms and ferns, should be photographed intact and next to a scale before they are cut into portions for pressing. A flash or tripod is needed for most shots inside the forest. When it is not possible to carry the camera to the field (during rainy weather or when collecting alone, for example), collected bulky plant parts may be photographed in the base camp.

MEASURING TAPE This is used to measure the diameter at breast height (dbh) of trees as well as large leaves and inflorescences that need to be cut up before they are pressed. Some forester tapes have both metric and dbh scales.

POCKET HAND LENS (10×) This is used as an aid in identifying specimens in the field.

MASKING AND DUCT TAPE Masking tape can be used to label specimens, tie packages, and so on. Duct tape is very useful for general repairs of equipment.

TREE CLIMBERS AND HARNESS These are needed when herbarium collections must be made from trees whose canopy is inaccessible from the ground. *Tree climbing is dangerous and should be done only by trained personnel.* In many areas, it is possible to hire local tree climbers. There are types of equipment available for climbing trees, including French telephone pole climbing spikes or "griffes," single forester spikes, climbing straps or *peconhas,* "Swiss Tree Grippers" (Mori, 1984, 1987), and mechanical ascenders (Perry, 1978; Perry & Williams, 1981; Whitacre, 1981).

Using mechanical ascenders requires the most amount of equipment, expertise, and time and is usually appropriate only for longer term studies, when repeated access to one tree is desired. French telephone pole climbing spikes ("griffes") provide the cheapest and lightest method of access, but they are suitable only for trees with a diameter of 49 cm or less. The canopy of larger trees can often be accessed from smaller trees using the extendable tree pruner. Forester spikes are also lightweight and cheap, and they can be used on larger trees, but they do not provide the level of support of griffes. Both types of spikes injure trees and probably increase mortality, so their use may not be acceptable in certain areas or in long-term studies. The Swiss Tree Grippers are usable for diameters from 18 to 72 cm and, unlike spikes and griffes, do not damage the tree. They are, however, considerably heavier and more expensive.

Specimens from trees may also be collected by throwing a line over a small branch and then either pulling until the branch breaks (Hyland, 1972) or sending a wire saw or cutter over the branch to cut it down (Collis & Harris, 1973). A review of techniques used to climb and collect specimens from trees may be found in Wendt, 1986.

FIRST-AID KIT   Collectors should be able to provide first-aid treatment for broken or sprained limbs, insect and snake bites, allergic rashes, cuts, blisters, minor infections, and sunburn.

General Considerations

Choosing the Sample

Most plant individuals will have a range of foliage and inflorescence sizes. The sample chosen for pressing should be representative of the size, variation, and general appearance of the plant or population collected. Even though collection of excessively damaged leaves and plant parts should be avoided, the selection of samples for pressing should not be based on their attractive appearance or convenient size but rather on their representativeness of the plant or species in question.

Gathering Sufficient Material

Enough material should be gathered to produce as many duplicates as needed. Usually at least three sets are needed: one set for the collector's institution, one set for the specialist determining the specimen, and one more set for the host country. Additional sets are often required or welcomed by other institutions in the host country and abroad. Thus, most botanists collect from five to eight fertile duplicates for each collection number. Three duplicates of sterile specimens, however, is an adequate amount. Collectors should gather more than enough flowers and fruits; some may become dislodged or damaged, and taxonomists may need to remove and dissect a few. Extra flowers may also be packed in small packages and included with the specimen, as these are often extremely useful for identification. Bulky inflorescences may be trimmed down, and the cuttings can be dried separately and eventually included in a small package. Enough individuals of small plant species should be included to fill a whole sheet, but collectors should ensure that all individuals are from the same population. Also, care should be exercised, particularly when collecting in frequently visited areas, not to deplete the populations of rarer species.

In some cases, flowers are taxonomically more important than fruits, and in some cases the opposite is true. Ideally, both should be collected, so two collections may be required: one during flowering and another during fruiting. For trees, it may be useful to include a strip of bark or a picture of a small slash made on the trunk with a knife. Collections of small plants should include the roots, but care should be taken to remove all soil.

Sterile versus Fertile Specimens

Sterile specimens, without flowers or fruits, are difficult to identify and are taxonomically useless. Their determination is at best unreliable and usually is incomplete. Sterile collections should be made only if the plant is important from an ethnobotanical standpoint and if a fertile specimen cannot be located. Efforts should be directed during the remaining time in the field to collect a fertile specimen. Individual trees can be marked with a plastic ribbon and collected at a later date when they are fertile.

General Procedure

Using a Field Press

As stated previously, specimens should be pressed in the field whenever possible. Before pressing herbaceous plants, shake the roots or wash them to remove any mud or sand. Clumps of grasses and sedges may need to be broken up, in which case the fact that clumps were broken should be indicated in the notes. Specimens should be cut and/or folded to fit neatly inside a folded sheet of newspaper and then slipped between two pieces of felt paper and placed inside the field press.

Whenever possible, the whole plant should be collected. Some herbaceous plants and most woody plants are too large to fit in a newspaper sheet, even when folded. In this case, an appropriate section of the plant must be selected for the press. A plant with intermediate-sized foliage and inflorescence should be selected, together with a separate leaf of the largest size found on the plant. All duplicates should be collected from the same individual, except for individuals too small to fill a herbarium sheet. In this case, collectors can include several individuals of the same population.

Large compound leaves, such as those of palms and ferns, should be measured and the number of pinnae or leaflets counted. Photographs of the whole leaf and plant are also useful. Apical, mid, and basal sections of the appropriate size can then be cut (see Balick, Chapter 5, this volume). If not pressed in the field, loose parts should be tagged and carried in a collecting bag back to the base camp, where they can be dried separately.

Delicate or ephemeral flowers should be placed in small envelopes of folded, preferably waxed, paper with some corollas open to show the internal structures. These envelopes can be placed with the plant specimen or labeled with the same collection number and carried separately in a pocket press or notebook.

Using a Collection Bag

When carrying specimens in a bag for subsequent pressing in the base camp, collectors should cut each sample to a size slightly larger than that of the pressed specimen. Each sample should be labeled with masking tape or a paper tag with its respective collection number. This step is very important, as collections can easily get mixed up. All samples of the same collection should be carefully bundled and gently but firmly pushed to the bottom of the bag.

Delicate plants or plant parts, as well as loose material, should be placed separately in labeled bags or enclosed in a newspaper sheet and placed in the field press. Large, heavy fruits or samples should not be placed on top of the bag, as they will crush the specimens underneath. Instead, they should be labeled and packed in a separate bag or placed at the bottom of the collecting bag.

To prevent plants from wilting, do not leave specimens in the bag for too long, especially in the heat. The bag should be kept in a cool, shaded spot until the material can be pressed. Transparent plastic bags should be avoided unless placed inside an opaque bag, as the "greenhouse effect" will accelerate wilting of plant specimens. In addition, wet newspapers can be placed inside the bag, or water can be sprinkled over the specimens.

When ready to press, do not pull specimens out of the bag, because they will tear. Instead, hold the bag upside down and carefully shake out its contents while trailing the bag across the floor or tarp. This method should leave a neat row of plant bundles along the floor, each corresponding to a different collection.

Collection Numbers

A standardized numbering system should always be used to label all collections and cross-reference them with the field notes. Every collector should have his or her own collection number series, which usually consists of his or her name followed by a number. Collections should be numbered sequentially, beginning with number 1 and proceeding ad infinitum. Newspaper sheets containing specimens should also be labeled with the collection number: many collectors also include their initials to distinguish them from the numbers of other collectors. Numbering systems that initiate a new number series every field trip or year are *not* recommended because repeated numbers may subsequently be confused and will create difficulties for workers and monographers referring to the collections.

The following points should be borne in mind:

- Any one collection number should be used only once. If two collections are accidentally given the same collection number, they can subsequently be distinguished by using a letter as a suffix (e.g., *a* and *b*). Thus, all duplicates of Martin Smith's first collection will be M. Smith 1 or MS 1. There should never be another Martin Smith collection with the same collection number.
- Except for very small plants, in which case one individual is insufficient to fill one sheet, any one collection number (including all its duplicates) should contain fragments from **only one individual plant.** Other individuals, even if they are of the same species and variety, should be given new collection numbers. Similarly, if the same individual plant is collected at a later date, a different collection number should be used.
- In cases where each set consists of several sheets, as with large-leafed plants, each sheet should be labeled accordingly: for example, sheet 1 of 3, sheet 2 of 3, etc. When a group of collectors are collecting together, only one collection series number should be used for each plant: the same individual plant *should not* be collected using several collectors' numbers.

The Collection Notebook and Field Notes

Field notes are one of the most important, yet most frequently overlooked, aspects of collecting. Good field notes make the herbarium sheet a valuable research tool and are essential for obtaining accurate taxonomic determinations. Notes should be taken when collections are made, since later in the day it is easy to forget or confuse important details of the plant or collecting site. Notes should be written *clearly* and *legibly,* especially if they are to be transcribed onto labels by others. Abbreviations should generally be avoided, unless these are universally recognized. If abbreviations are used, a key must be provided in the notebook.

Some botanists prefer to take notes in the field in a small pocket book, transcribing the final notes neatly onto a larger book at the end of the day. This procedure takes more time, but the choice of method is up to the individual. Collectors *should not* wait to get back to the camp to write field notes from memory. Also, a separate set of photocopies or carbon copies of the collection notebook should always be kept, in case the notes are lost, stolen, or damaged.

In addition to the collection number, date, and collectors' names, the following information should be included in the collection notebook for each collected plant.

- Latitude and longitude. Can be obtained from a detailed map of the area.
- Altitude above sea level. Can be obtained from large-scale topographic maps or, preferably, read from a calibrated altimeter. Although extreme accuracy is unnecessary, some of the cheaper makes should be avoided, as these may be highly inaccurate and subject to variation.
- Locality. Country, state or region, province.
- Exact location. The description of the exact location should be detailed enough to allow another person to find the site on a map or to return to the site at a later date. The distance and direction from important, well-known topographical or man-made reference points such as a river mouth, mountain top, town, or road should always be indicated.
- Habitat and vegetation. Type of habitat (e.g., riparian, primary forest, secondary forest, cultivated field, etc.) should be described. If possible, a description of the dominant vegetation, as well as of typical or associated species should be provided.
- Topography and soil. Notes should be included on the characteristics of the terrain (e.g., rugged, sloping, flat), as well as on the type of exposure (sunny, shaded) and soil type (e.g., clay, sand, alluvial, volcanic, well-weathered, extent of organic accumulations).
- Taxonomic identification (family, genus, species). Field identifications are only tentative, especially beyond the level of family. A list of field guides to neotropical families is provided in Appendix 1.
- Plant description. This should include information about characteristics of the plant not visible in the dried specimen. Some important characters include:
  Habit. Whether the plant is a tree, shrub, epiphyte, vine, herb (annual, biennial or perennial), solitary, or colonial.
  Height of plant. A rough estimate is usually adequate.

Stem or trunk characteristics. The diameter at breast height (dbh) of the stem or trunk of woody plants should be indicated. Any distinguishing characteristics of the bark, including color, texture and thickness, should also be noted. For fleshy stems particularly, collectors should indicate if the cross section is round, flattened, or square.

Architecture. Any characteristic type of branching (erect, pendulous, etc.) and crown shape should be noted.

Flowers and fruits. Color, odor, shape, and size should be indicated. The numbers of stamens and styles on flowers are easier to determine in the field than when the material is dry. Floral formulae are a useful and quick way of recording the number of parts. Fruits should be opened, and the aril and seeds described if present. Information on fruit texture and consistency should also be recorded.

Other. Any other characteristic anatomical or morphological feature that may not be visible in the voucher should be noted, including extremes of morphological variation found in leaves and other organs. The presence, consistency, and color of any **latex, resin,** or **sap** should be indicated, noting any differences between trunk and twig exudate. Finally, the presence of any insects associated with the plant, including pollinators and symbionts, should also be recorded.

- Number of duplicates collected. This can be annotated under the collection number, in the left-hand margin of the field-book. This information is necessary to print the correct number of labels during processing of collections at a later stage. The number of duplicates left in the host country should be specified for each number, so that labels can be sent there.

- Others. Collectors should also indicate in their field notes whether any photographs have been taken, whether fruits have been dried and stored separately, and whether the vouchers were preserved in ethanol or given any other chemical treatment.

- Local name and uses. See Alexiades, Chapter 3, this volume.

To save space and time, you can note the date, collectors, latitude and longitude, altitude, and locality at the beginning of a series of collection numbers, if they are all collected in the same area on the same date and by the same people. A change in collecting date, collectors, or geographical location should mark the beginning of a new section (but not a new series of numbers) in the collection notebook. Some collectors indicate at the start of each series which collection numbers are included for that particular series. Another system is to write the full information for the first collection number (e.g., 509). If the next 10 collections were made in the same locality, "509" can be written next to "Locality" for each. The latter system is generally more cumbersome, especially when typing labels from the notes or when needing to refer to the collection notebook. The beginning of a new series of collections is often indicated by drawing a horizontal line across the page. The increasing availability of portable computers makes electronic data entry an increasingly accessible option.

The Herbarium Label

The herbarium label contains the information recorded in the field collection notebook and is affixed to the herbarium specimen. Although labels can be mechanically typed on any acid-free, archival paper, the most effective way to prepare labels is by using a label-generating program, usually as part of a database or word-processing software. The advent of portable hardware means that labels can be produced in the field, thus speeding up the processing of collections. A typical layout of information and sample herbarium labels are illustrated in Figures 3 and 4.

Pressing and Drying
Plant Specimens

Plants collected in a plastic bag should be pressed as soon as possible. Plants brought from the field press should be transferred to an ordinary press for drying.

The Plant Press

Plant presses are used to dry plant specimens while keeping them flat. Freshly cut plant specimens are arranged inside a folded

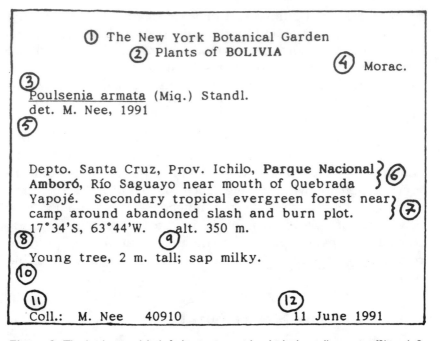

Figure 3. The herbarium label. **1.** Institution with which the collector is affiliated. **2.** Project title. **3.** Genus, species, and author. **4.** Family. **5.** Specialist and date of determination. **6.** Locality. **7.** Vegetation and habitat. **8.** Latitude, longitude. **9.** Altitude. **10.** Plant description. **11.** Collector(s). **12.** Collecting date.

sheet of newspaper, which is in turn sandwiched between two sheets of blotting paper and two pieces of corrugated cardboard or aluminum, each 12 × 18 in. (Figure 5).

After all specimens have been arranged on top of each other and sandwiched between blotters and corrugated sheets, a wooden frame is placed at both extremes of the pile and tied together as tightly as possible with a pair of ropes or webbing straps, 6–10 ft long. The blotting papers gradually absorb the water from the plant tissues, while the pressure on the specimen prevents the specimen from wrinkling. The corrugated sheets facilitate the flow of air through blotters, thereby assisting in the drying of the plant specimen.

Plant press frames may be purchased, cut from plywood, or made from slats (¹/₄ in. thick, 1 in. wide) of any hard, flexible wood. Five pieces, 18 in. long, and six pieces, 12 in. long, are nailed or riveted together to make a lattice. All the longer pieces

ECONOMIC BOTANY OF THE BORA INDIANS: PERU

No. 1006

Abuta rufescens Aublet

det. B.A. Krukoff 1981

Dept. Loreto, Prov. Maynas: Rio Yaguasyacu, affluent of Rio Ampiyacu.
Brillo Nuevo and vicinity, approx. lat. 2°40', long. 72°00'

Vine 2 meters tall, leaves dark green above, light green beneath. Underside of leaves and stem are hairy. In 1° forest.

Bora: namityahkeu

Use: to make curare, now little used as shotguns are common.

M.J.Balick,D.R.Allon,J.P.Razon   2/78

The New York Botanical Garden
INSTITUTE OF ECONOMIC BOTANY
Tambopata Ethnobotanical Survey Project

MALVACEAE

Gossypium

PERU, Dpto Madre de Dios. Provincia de Tambopata. Rio Tambopata. Comunidad Nativa de Infierno. Quebrada de Torén. 12°50'S, 069°17'W. 00260m. Cultivated next to house of Don Walter Torén.

Shrub to 2m, branched; flowers yellow, maturing pinkish.

n.v.: Algodón Blanco [Spanish].
USE: Antifungal for skin fungus (unripe fruit). Apply fruit
    peel scrapings
    Parturifacient (leaves). Drink infusion
M. N. Alexiades 1006   August 1, 1990

Fieldwork Supported by the Edward John Noble Foundation

Figure 4. Samples of herbarium labels.

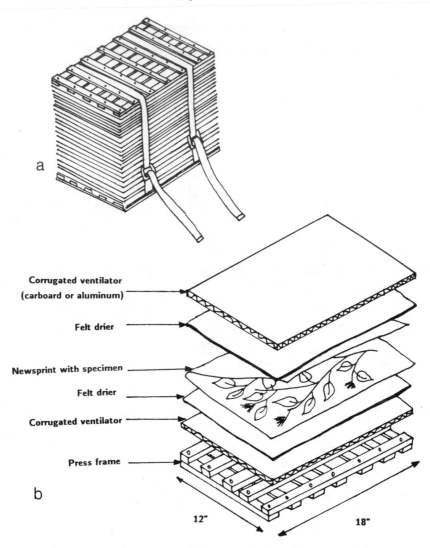

Figure 5. The plant press. **a.** Assembled view. **b.** Exploded view showing components. (Redrawn from Savile, 1962.)

should be placed on the same side of the lattice. Plant presses can also be improvised in the field by placing plants between two boards and applying pressure with a heavy rock.

Blotters and corrugated sheets may be purchased, though it is also possible to use thick pads of newspaper instead of blotters, and corrugated sheets can be cut from cardboard boxes (be sure that the channels run along the width of the board). Cardboard blotters are cheap and lightweight, but they need to be dried at regular intervals because they absorb moisture from the drying specimens. Also, time and use cause the corrugations to collapse, and the sheets need to be replaced. Aluminum blotters, on the other hand, are very durable but heavy and expensive. Because cardboard corrugated sheets need to be regularly taken out of the press and dried, collectors need to carry more of them than aluminum ones.

While specimens are drying, particularly during the early stages, the press should be regularly checked to ensure that the straps remain pulled tightly enough: as the plant specimens dry and shrink, the press tends to become loose, resulting in wrinkled specimens.

Arranging Specimens for Pressing

Correctly arranging the specimen in the newspaper before drying is very important; dry specimens are brittle, and their parts cannot be rearranged without breaking off. When arranging the specimen, collectors should ensure that it is cut to the right size. To do so, place it over the newspaper and cut the excess length from the base (Figure 6). When cutting off leaves or twigs, leave a portion of the base intact to indicate its position. It is also possible to fold the stem (Figure 7).

Both surfaces of the leaves and reproductive structures should be visible, so at least one leaf and one flower should be turned over. If only one leaf fits in the newspaper, it should be folded in such a way that both surfaces are exposed (see Figure 1). Collectors should note that often the lower leaf surface has more diagnostic characters than the upper surface. If necessary, one side of very large symmetrical leaves may be removed.

Some flowers should be pressed open, some closed. If possible, one flower should be dissected to show the internal structures. Extra fertile material for study and dissection should also be included whenever possible. Care should be taken that nodes, flowers, or fruits are not accidentally covered by folded leaves. Large leaves should be folded away from these parts or tucked underneath (Figure 8).

Leaves should be folded in a way that exposes most of their

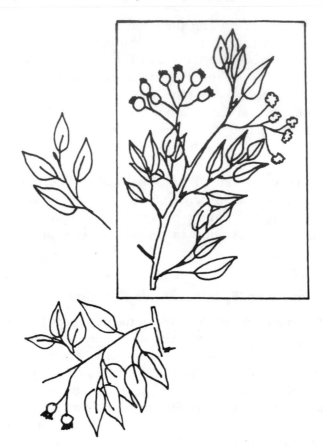

Figure 6. Cutting the specimen to size. (Redrawn from Liesner, 1985.)

surface—especially the apex and base (Figure 9). Leave all the leaves you can; trim only if they lie several deep on the sheet. Do not leave any plant part projecting out of the newspaper; these parts will break off when they are dry. Instead, fold or tuck in neatly any leaves or stems that jut out of the news sheet.

Pressing Large or Fleshy Parts

Sections from large tubers, corms, bulbs, and the like, may be cut, noting the thickness and length of the part. Whenever possible, the outside surface should be included. Bulky adventitious root systems can also be thinned down. Whenever pressing samples with bulky parts, pad hollow spaces with cardboard or foam

Figure 7. Folding the specimen. (Redrawn from Liesner, 1985.)

to maintain a constant pressure. This padding prevents corrugated sheets from slipping out from the bottom of the press when pressure is applied, and it also keeps specimens from wrinkling.

Cross and longitudinal sections of fruits should be included whenever possible. Fruits can also be cut and dried in a separate sheet and combined with the leaves once dry. Large fruits, particularly when hard, can also be dried in individual nylon net,

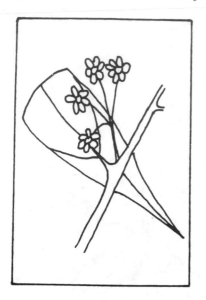 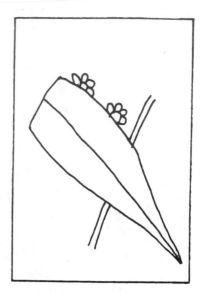

YES !   NO !

Figure 8. Correct and incorrect ways of folding leaves. (Redrawn from Liesner, 1985.)

cotton, or (least preferable) paper bags. When fruits are separated from the main collection, the caption "Fruits Separate" should be written in the newspaper folder and in the collection notebook. Large inflorescences can be photographed, measured, and thinned. Many of these thinnings can be pressed separately and packed as extra material for dissection.

Balancing the Press

Usually no more than 60 to 90 specimens, fewer if they include bulky material, should be placed in a single press; larger presses often become unwieldy and difficult to balance. When building a press, keep the numbered side of sheets facing up. The press can be balanced by turning alternate collections around, so that the open ends of the folded newspaper face opposite directions. The press can also be balanced by placing large fruits and stems on different parts of the sheet, padding these with pieces of foam.

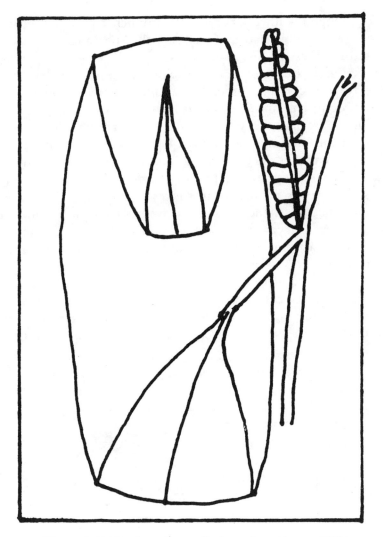

Figure 9. Folding large leaves. (Redrawn from Liesner, 1985.)

Drying Specimens

Plants kept for too long before drying may mold, and the leaves may abscise. It is possible to dry small numbers of specimens by leaving the press in the sun, replacing the moist blotters with dry ones every 8 hours. The moist blotters and corrugated cardboard should be thoroughly dried in the sun and the process repeated until the specimens are dry. No attempt should be made at this time to change the newspaper sheets in which the plant specimen was pressed, as doing so may damage the specimen and lead to confused numbers or lost plant parts.

An artificial source of heat is usually necessary to dry specimens. Presses can be suspended above stoves or open fires, placed close to chimneys or next to generator exhausts, . . . the possibilities are innumerable. In any case, care should be exercised to ensure that there is enough heat to dry the plants as fast as possible while, at the time, avoiding excessively high temperatures, which could cause a fire or destroy the specimens: the press should never be too hot to touch.

One of the safest, most common ways of drying plants is to build a dryer. A dryer consists of a wooden or aluminum rectangular box measuring 18½ in. across, 23 in. high, any desired length, and a heat source. Two cleats, about 1 in. wide, run along the top side of the box supporting the plant press. A wire screen is often stretched between these cleats to prevent anything from falling down onto the heating elements and starting a fire. A prototype for a portable dryer is shown in Figure 10. A higher

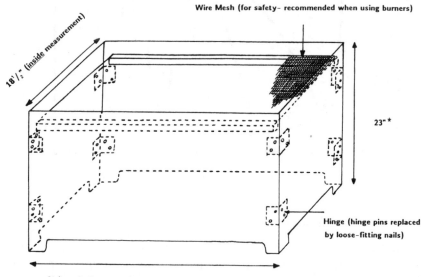

Figure 10. Portable plant dryer. *=A longer section, totally or partially enclosing the plant press, is more efficient but bulkier. **=This section may be cut into two or more parts and joined by hinges, making it more portable.

box in which the plant press is totally or partially enclosed is more efficient but obviously bulkier and heavier.

When electricity is available, six or more 100- or 200-watt light bulbs can be used as a heat source. Light bulbs are safe, but they need constant replacement and in some areas are expensive. Heat lamps, resistance coils, or heating strips may also be used. In the absence of electricity, bottled gas provides a good source of energy. Portable kerosene stoves sold all over Latin America are very basic, cheap, and effective. Any naked flame should be covered with a metal plate in order to minimize the risk of fire and to diffuse the heat. When drying specimens, use a lid to cover parts along the top length of the dryer that are not covered by the plant press; this arrangement will force hot air through the press corrugates.

Excessive drying of specimens will "cook" them and make them very brittle. To check if specimens are dry, feel the fleshier parts, such as stems and fruits, as these take the longest to dry. One way of checking is to press gently with a fingernail until a slight crackle is heard or felt. If any parts feel cool, soft, or damp, the specimen is not dry. Drying time will vary according to the amount of heat applied and the amount of water in the plant tissues; times vary from 8 hours to several days.

Preserving Plants before Drying

In some cases, it is not possible to press and dry plants in the field. One possibility is to soak bundles of plant specimens with alcohol inside heavy-duty plastic fertilizer bags until they can be dried. Specimens can be stored in good-quality plastic bags with 70% or higher alcohol solution for several weeks. Although more-dilute solutions can be used for shorter-term storage, solutions of less than 40% should be avoided. Some botanists add formaldehyde to this solution, particularly as this seems to subsequently offer some protection to dried specimens against insect attack. But formaldehyde makes subsequent handling of specimens unpleasant and potentially hazardous owing to toxic fumes.

Specimens to be kept in a preserving solution should be packed into package bundles 9 in. thick, wrapped in newspaper sheets, and packed snugly into the plastic bag. The alcohol, 1–2 pints per bundle, should then be poured over the bundles. The use of any preserving solution should be clearly indicated in the collection book and on the specimen labels, since some of the properties of the specimens, including color and chemistry, may be altered by the procedure. *Note: When preserving specimens in alcohol, test the pen to ensure that its markings on the newspaper sheets will not be erased.* Wax crayons are a good writing instrument when specimens are to be preserved in alcohol.

Packing and Shipping Specimens

Dry specimens should be arranged into 9–12 in. bundles, placed between two cardboard sheets, and securely tied with heavy twine. For additional protection, each bundle can be wrapped in heavy paper or plastic. Naphthalene may be added to prevent insect damage. The bundles should be stored in a dry place. Under humid conditions, it may be necessary to pack bundles in tightly sealed fertilizer bags. When shipping specimens, snugly pack bundles in strong wood or corrugated cardboard boxes. The latter may not be suitable if the boxes are to be exposed to rain or humid conditions. Collectors should arrange the necessary export and import permits before shipping any specimens. Often, clearance from CITES (Convention on International Trade in Endgangered Species of Wild Fauna and Flora) must be obtained, as well as a phytosanitary permit, certifying that specimens do not include endangered species and are not contaminated with mold or insect pests.

Acknowledgments

Camilo Díaz introduced me to the science and art of collecting plants. Bradley Bennett, Scott Mori, Michael Nee, Christine Padoch, Daniela Peluso, Charles Peters, Oliver Phillips, and Jennie Wood Sheldon provided many helpful comments on earlier drafts of this paper. I am also grateful to Michael Balick and Jan Stevenson for their helpful suggestions and support in writing these guidelines. The Institute of Economic Botany of The New York Botanical Garden and the National Cancer Institute supported the preparation of this paper.

Literature Cited

Archer, W. A. 1945. Collecting data and specimens for study of economic plants. Miscellaneous Publication No. 568. U.S. Dept. of Agriculture, Washington, D.C.

Bye, R. A., Jr. 1986. Voucher specimens in ethnobiological studies and publications. Journal of Ethnobiology 6(1): 1–8.

Collis, D. G. & J. W. E. Harris. 1973. Line-throwing gun and cutter for obtaining branches from tree crowns. Canadian Journal of Forestry Research 3(1): 149–154.

De Wolf, G. P. 1968. Notes on making an herbarium. Arnoldia 28: 69–111.

Fidalgo, O. & V. L. Ramos B (eds.). 1989. Técnicas de coleta, preservaçâo e herborizaçâo de material botânico. Instituto de Botânica, Secretaria do Meio Ambiente, Governo do Estado de Sâo Paulo, Sâo Paulo.

Forman, L. & D. Bridson (eds.). 1989. The herbarium handbook. Royal Botanic Gardens, Kew.

Fosberg, F. R. 1939. Plant collecting manual for field anthropologists. American Fiber-Velope, Philadelphia.

——— & M. H. Sachet. 1965. Manual for tropical herbaria. Regnum Vegetabile 39. International Bureau for Plant Taxonomy and Nomenclature, Utrecht, Netherlands.

Hyland, B. P. M. 1972. A technique for collecting botanical specimens in rain forest. Flora Malesiana Bulletin 26: 2038–2040.

Johnston, I. M. 1939. The preparation of botanical specimens for the herbarium. Arnold Arboretum, Jamaica Plains, Mass.

Ketchledge, E. H. [1970?] Plant collecting: A guide to the preparation of a plant collection. State University College of Forestry at Syracuse University, Syracuse, N.Y.

Liesner, R. 1985. Field techniques used by Missouri Botanical Garden. Compiled by Ron Liesner with suggestions from the staff. Unpublished manuscript.

Lot, A. & F. Chiang (comps.). 1986. Manual de herbario. Administración y manejo de colecciones, técnicas de recolección, y preparación de ejemplares botánicos. Consejo Nacional de la Flora de México, A.C., México.

Mori, S. A. 1984. Use of "Swiss Tree Grippers" for making botanical collections of tropical trees. Biotropica 16: 79–80.

———. 1987. The Lecythidaceae of a lowland neotropical forest: La Fumeé Mountain, French Guiana. Memoirs of The New York Botanical Garden 44: 3–7.

———, L. A. Mattos Silva, G. Lisboa & L. Coradin. 1989. Manual de manejo do herbário fanerogâmico. 2ᴬ ed. Centro de Pesquias do Cacau, Ilheus, Brasil.

Perry, D. R. 1978. A method of access into the crowns of emergent and canopy trees. Biotropica 10: 155–157.

——— & J. Williams. 1981. The tropical rain forest canopy: A method providing total access. Biotropica 13: 283–285.

Savile, D. B. O. 1962. Collection and care of botanical specimens. March

1962. Reprinted with Addenda 1973. Publication 1113. Research Branch, Canada Department of Agriculture, Ottawa.

Smith, E. C. 1971. Preparing herbarium specimens of vascular plants. Agriculture Information Bulletin No. 348. Agricultural Research Service, U.S. Dept. of Agriculture, Washington, D.C.

Wendt, T. 1986. Arboles. Pages 133–142 in A. Lot & F. Chiang, comps., Manual de herbario. Administración y manejo de colecciones, técnicas de recolección, y preparación de ejemplares botánicos. Consejo Nacional de la Flora de México, A.C., México.

Whitacre, D. F. 1981. Additional techniques and safety hints for climbing tall trees, and some equipment and information sources. Biotropica **13:** 286–291.

Womersley, J. S. 1981. Plant collecting and herbarium development: A manual. FAO Plant Production and Protection Paper 33. Food and Agriculture Organization of the United Nations, Rome.

5

Collecting Palm Specimens

Michael J. Balick
*Institute of Economic Botany,
The New York Botanical Garden*

Introduction

Palms are distributed throughout much of the tropics, including the Neotropics. Economically and ethnobotanically, palms are one of the most important plant groups of the tropical forest (Balick & Beck, 1990). In some areas, they are scattered throughout the vegetation; in others, palms are the dominant forest trees. In fact, oligarchic stands of palms are quite common, especially in habitats that are not well tolerated by most dicotyledonous species, such as swamps and areas subject to seasonal inundation (Peters et al., 1989). Despite their relative abundance and impor-

Selected Guidelines for Ethnobotanical Research: A Field Manual, 127–133
Edited by Miguel N. Alexiades
© 1996 The New York Botanical Garden

tance as a part of the overall tropical rainforest biota (Kahn & Granville, 1992), as well as their many economic and subsistence uses (Balick, 1988), palms are usually poorly represented in the study collections used by biologists. One study (Balick et al., 1982) documented the lack of good palm collections from one important region: only 37.5% of the 232 recognized palm species in Brazilian Amazonia were present in any of the three local herbaria within that zone. Of the species present, many lacked crucial information on their labels, and others were badly preserved. Although a surge of interest in and field studies of the palms have increased the numbers of specimens in local and international herbaria since that study, the group is still underrepresented. As a result, local studies of ethnobotany, ecology, population biology, and floristics are hindered by not having adequate and complete reference collections with which to compare material being brought in from the field.

It is important to emphasize that good collections of these plants are necessary in order to make positive identification possible. A poor palm collection is worse than no collection at all in that it tends to give a misleading impression of the plant or creates many unanswerable questions. Because palms are often rather large, or contain organs that are hard to press, collecting them often requires a considerable effort.

Collection Techniques and Methods

Here I summarize some of the techniques and methods for making a good palm collection. This information is presented in greater detail in Balick et al., 1982, and the interested reader is also directed to Fosberg & Sachet, 1965, for additional information on this topic.

Because of the rather large size of many palm parts, it is difficult to collect entire organs such as leaves, which may be up to 8 m long and may weigh 20 kg or more when fresh. Similarly, a fruiting panicle with 30 kg of coconut-like fruits would be extremely difficult to collect, dry, ship, and curate in its entirety. Therefore, it is advisable to make collections with representative parts of each of the major organs. Through the use of photographs and very detailed measurements, one can subsequently reconstruct in the herbarium the general nature and habit of the palm. Good palm collections should include the following 10 elements:

1. *Leaf apex in pinnate-leaved palms; leaf hastula in palmate- or costapalmate-leaved palms.* The apical pinnae can be fused in a distinctive way or have dimensions altogether different from the middle pinnae. The hastula, or ligula, as it is sometimes known, often varies in shape and size and can provide some element of distinction at the species level.

2. *Leaf pinnae and at least a portion of the rachis in pinnate-leaved palms; leaf segments in palmate-leaved palms.* The pinnae are often crucial for making generic and specific determinations.

3. *Flowers, fruits, or both.* Flowers and, to a lesser degree, fruits usually contain the most essential criteria for systematic identification of palms, at both the generic and specific levels. It is essential that collections include such fertile material. It is often possible to locate an individual plant in flower while at the same time locating a similar individual of the same species having fruit in some degree of maturation. While it is preferable to obtain flowering and fruiting material from the same tree, it is also permissible to collect flowering and fruiting material from different individuals, if great caution is taken to collect from the same species. The herbarium label should note that sexual material was taken from different trees.

4. *Flowering or fruiting axes or both.* The morphology of the flowering or fruiting axes often provides important criteria for taxonomic identification. Smaller axes should be collected in their entirety; larger ones should be photographed and measured carefully.

5. *Supplementary material: bracts, sheaths, spines, wood samples, stems, and seedlings.* Such materials may be helpful in identification, depending on the taxon in question.

6. *Quantitative and qualitative information on vegetative, reproductive, and morphological characters not apparent on the specimen.* It is essential to spend as much time as needed to properly document the presence, distribution, and dimensions of vegetative, reproductive, and morphological characters. In herbarium collections, there are many important

bits of information that cannot be fully and accurately preserved or discerned. For example, some palm fruits
change color when dry. Another example would be the
dimensions of a large leaf, which might be represented in
the herbarium collection by only a sample of the apical,
middle, and basal pinnae and rachis. It is very important
that proper measurements be made in the field and faith-
fully transcribed on the herbarium label.

7. *Information on habitat.* These data should include comments
   on altitude, substrate, vegetation type, and degree of habitat disturbance. Because many species of palm are limited
   to specific habitats, such ecological indicators may be especially useful in identifying a species.

8. *Reasonably precise locale.* As is the case in all herbarium collections, information on locale is essential, especially if the
   specimen represents a newly discovered, rare, or endemic
   species.

9. *Vernacular names and uses.* Most palm species are referred
   to by vernacular names, and many have local uses. Vernacular names are extremely helpful in relocating species
   populations in the field, and ethnobotanical uses may be
   indicative of species of potential economic value worth developing further. It is surprising how many palm collections are lacking such information (see also Alexiades,
   Chapter 3, this volume).

10. *Good-quality black-and-white or color photographs.* Even if
    they are not contained in the herbarium collection, it is
    extremely important that the collector take adequate pic-
    tures of the palm before and after it is collected. Especially
    helpful, for example, are habitat photographs, a shot of
    the individual specimen(s) collected for harvest, and close-
    ups of the reproductive parts, leaf, stem, and other subsid-
    iary organs. These photographs are often kept by the col-
    lector, and their availability is indicated on the label.
    Although useful, it is admittedly rather expensive to in-
    clude original photographs in each herbarium collection.

Because the proper collection of a palm often results in sacrifice of the tree, it is important to make as many specimens as
possible from that particular individual. In general, I find that six
specimens are sufficient—from one to three deposited in the

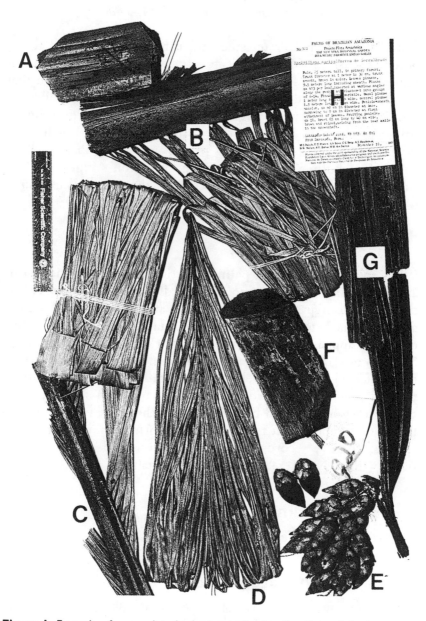

Figure 1. Example of a complete herbariam collection. Specimen of *Attalea maripa* (Aubl.) Mart., collected along the Santarém–Cuiabá Road, BR 163, Km 890 from Santarém, Pará, 10 Nov 1977, *M. J. Balick et al. 920*. Portions represented are: **A.** Section of Petiole. **B.** Lowermost section of leaf rachis with basal pinnae. **C.** Midsection of leaf rachis with folded pinnae (note that pinnae on left side of rachis are removed and only their bases remain to indicate position of insertion into the rachis). **D.** Apex of leaf. **E.** Section of fruiting panicle, several rachillae and sample of fruit. **F.** Stem section. **G.** Inflorescence bracts, sliced along their length. **H.** Herbarium label. (Originally published in *Brittonia* 34: 470. 1982.)

country of origin as requested and three duplicates sent to the home herbarium. More duplicates can be made, although drying these in the field is somewhat difficult, especially if more than one or two good collections are made each day and the collecting activity lasts for several weeks or more. It is not uncommon for palm collectors to find that this material has filled up the field dryers and takes many days to process, resulting in a potential backlog of both palm and nonpalm collections. Figure 1 illustrates a complete herbarium collection of the palm *Attalea maripa* (Aubl.) Mart. from Brazil.

Conclusion

While seemingly difficult, palm collection becomes an interesting and rewarding endeavor once the hurdle of the first few collections has been passed. The collector should be armed with a good ax, machete, clippers, measuring tape, 35-mm camera, and copious supplies of color slide and black-and-white print film. Strong cord and woven plastic sacks are important for carrying the collected material back to the location where it is to be pressed and otherwise processed. The fact that these plants have been somewhat ignored by collectors in the past gives an opportunity to provide great treasures in the form of new hybrids and species, country records, uses, distribution, and other information for today's generation of tropical botanists.

Acknowledgment

This is a modified version of Balick, 1989.

Literature Cited

Balick, M. J. (ed.). 1988. The Palm—Tree of life: Biology, utilization and conservation. Advances in Economic Botany **6.**

———. 1989. Collection and preparation of palm specimens. Pages 482–483 *in* D. G. Campbell & H. D. Hammond, eds., Floristic inventory of tropical countries: The current status of plant systematics, collections, and vegetation, plus recommendations for the future. The New York Botanical Garden, Bronx.

——— **& H. T. Beck (eds.).** 1990. Useful palms of the world: A synoptic bibliography. Columbia University Press, New York.

————, A. B. Anderson & M. F. da Silva. 1982. Palm taxonomy in Brazilian Amazônia: The state of systematic collections in regional herbaria. Brittonia 34: 463–477.

Fosberg, F. R. & M.-H. Sachet. 1965. Manual for tropical herbaria. Regnum Vegetabile 39: 1–132.

Kahn, F. & J. J. de Granville. 1992. Palms in forest ecosystems of Amazonia. Ecological Studies 95. Springer-Verlag, New York.

Peters, C. M., M. J. Balick, F. Kahn & A. B. Anderson. 1989. Oligarchic forests of economic plants in Amazonia: Utilization and conservation of an important tropical resource. Conservation Biology 3: 341–349.

6

Recommendations for Collecting Mushrooms

Roy E. Halling
Institute of Systematic Botany,
The New York Botanical Garden

Introduction

Mushrooms, fleshy fungi, are an important ethnobotanical resource in many societies (Fidalgo, 1965, 1968; Fidalgo & Hirata,

Selected Guidelines for Ethnobotanical Research: A Field Manual, 135–141
Edited by Miguel N. Alexiades
© 1996 The New York Botanical Garden

1979; Findlay, 1982; Wasson, 1975), particularly as sources of food (Beuchat, 1987; Parent & Theon, 1978; Prance, 1984) and in magical or religious contexts (Heim & Wasson, 1959, 1965; Riedlinger, 1990; Wasson, 1969, 1980). Although the techniques employed for collecting mushrooms are not complex, they differ considerably from those used to collect vascular plants. Pressing mushrooms will destroy their value as scientific specimens and often will render their taxonomic determination impossible. The following is a summary of techniques for gathering, documenting, and preserving mushroom specimens. Although these may seem at times idiosyncratic and tedious, they are essential for providing useful specimens for later study.

Collecting Specimens

Adequate notes are essential to make complete and valuable specimens. When collecting a mushroom, take note of its habitat and substrate. Common substrates for mushrooms include wood, soil, and leaf litter. If a mushroom is growing on wood, note whether the wood is dead or living. If the wood is living, is the mushroom growing on the bark, or is the wood decorticate? The habitat description should include the kind of trees growing in the area, as many agarics will associate with particular types of tree roots, or they may be substrate-dependent.

When collecting the specimen in soil, be sure to dig down deep enough to remove the whole agaric. Otherwise, remove part of the substrate with the specimen still attached. Try to collect young as well as mature specimens; several stages of development may be necessary for identification purposes. Furthermore, as many individuals as possible of one "taxon" should be collected.

Once collected, the mushroom must be handled carefully. Specimens should be wrapped in waxed paper (never in plastic!) in such a way that moisture cannot escape. A sheet of paper is torn so that a collection can be rolled up inside with the ends twisted closed. Waxed sandwich bags are also useful if available. The wrapped mushroom can then be placed in a sturdy basket or box, again never in a plastic bag, and carried to the laboratory. Never pile mushrooms so high on top of one another that delicate structures are broken. Small tin boxes or rigid plastic boxes can be useful for protecting fragile specimens.

Preparing Specimens

You should begin working on your collections as soon as possible after arriving from the field. Many agarics shrivel or fade within a few hours of collection, even when wrapped in waxed paper. Prevention of overheating and waterlogging during transport to the lab will aid in maintaining the specimens in as fresh a state as possible. It is best if collecting can be done in the morning with the remaining daylight or afternoon allocated to work on collections. You can expect to spend about 15–30 minutes per collection preparing spore prints and notes; a bit more time is required if photographs are taken.

Making Spore Prints

The first thing to do is to prepare spore prints. As the name indicates, spore prints are pieces of paper covered with agaric spore deposits. These are invaluable for identifying many specimens. To make a spore print, first remove the stipe, if present, and place the gill or pore surface down on a white (*never* black or colored) piece of paper. Cover the whole mushroom with some type of enclosed or moisture-resistant container (a drinking glass or jar or plastic sandwich bag; even wrapped in waxed paper can be sufficient). If there are only one or two specimens, it is better to cut a hole in the paper for the stipe rather than removing the stipe.

After one to several hours (sometimes overnight), a white or colored spore print should result (old or immature mushrooms may not give a spore print). Note the color of the fresh spore print, then fold and dry it with the specimen. In many cases, where return to the laboratory necessitates travel from one extreme elevation to another, it is better to attempt the spore deposits in the field. In my experience, agarics collected at high elevations and returned to low elevations will not sporulate. In these instances, however, one can facilitate spore deposits by placing the enclosed preparations in the bottom of the basket or box, even while still collecting, with an attendant note explaining to which collection the preparation belongs.

Making Notes

While the spore prints are being prepared, you can begin to take notes on your collections. Describing the fresh characteristics of agarics is of paramount importance to preparing a valuable specimen because many important and diagnostic features will disappear when the specimens are dry. Most importantly, color, shape, and size will change, and the odor or taste, if present, will no longer be evident. Notes on the fresh appearance should include any descriptive information that will not be evident after drying.

Essentially, agaric sporocarps can be divided into three parts: the **pileus,** or **cap,** including the interior flesh, the **hymenophore** (lamellae, or tubes or pores), and the **stipe** (including the interior flesh). Other features may, or may not, be present. These might include a **universal veil** and a **partial veil;** both are discussed in more detail below.

Pileus (Cap)

- Size range or diameter
- Shape, viewed and described as if sectioned longitudinally. Caps can be convex, concave, bell-shaped, mammillate, etc.
- Color. Center versus margin; surface ornamentation versus background; does color change with age or when the cap is bruised and handled?
- Texture and ornamentation. Is it hairy, smooth, scaly, fibrous, fragile, membranous? Is it slimy, dry, moist, sticky? Is the margin (outer edge) different from the center? What is the overall thickness?
- Odor and taste. *Never swallow a mushroom.* Masticate briefly, spit out, and note whether distinctive or not.

Hymenophore

- Type. Lamellae, or tubes or pores
- Color. Note changes between young and old or caused by injuries and bruises. If injured, is a juice or latex exuded? Is it colored? To what color does it change? Does it change slowly or rapidly? Does it stain surrounding tissues some other color?

- Attachment to stipe (when viewed in a longitudinal section from pileus down through stipe). Can be free, adnexed, adnate, decurrent. Ranges or intermediates may exist; use ranges not absolutes.
- Edge. Note color, any differences between the edge and the sides, and whether the edge is smooth or serrate.

Stipe

- Size. Include range of length and width.
- Shape: Can be equal, clavate, bulbous, tapering downward, etc.
- Attachment to pileus (e.g., central, eccentric, lateral, absent)
- Color when young and old, above and below, when handled or bruised
- Texture and ornamentation. Same as for pileus. Note basal mycelium and its color, abundance, etc.

Universal Veil

The universal veil is formed of tissue that completely surrounds the immature button stage of an agaric. It ruptures with stipe elongation and may leave remnants on the pileus surface or margin and on the stipe base or surface. It may be persistent or ephemeral. Remnants may appear as warts on the pileus and warts or concentric rings around the stipe base and on the stipe surface, or they may be flaplike patches on the pileus and a cuplike structure around the stipe base. As with other features, note colors and color changes.

Partial Veil

The partial veil is formed of tissue that extends from the pileus margin to the stipe and thus covers the hymenophore before maturity. It ruptures to form a ring around the stipe or a fringe of tissue at the pileus margin; intermediates may occur. Note its persistence, location, whether it is attached or movable. Again, note color and color changes, surface ornamentation, etc.

Drying Specimens

When you have finished your spore prints and have taken notes and photographs, the specimens can be dried. This is a very critical step and can make the difference between a valuable scientific specimen and a useless one. Dryers used for vascular plant specimens are usually too hot for drying mushrooms.

Methods for drying mushrooms vary from collector to collector and place to place. The feature they all have in common is that they utilize some kind of fine screen shelving, for suspending the mushrooms over a dry heat source. The heat source can be an electric heater, a tent heater, hot plate, light bulbs, kerosene lantern or stove, and so on. In any case, heat should be directed from the source upward, via a chimney effect, circulating around the specimens and escaping above. Also, it is critical that the specimens be dried slowly (not cooked) in a temperature of 55°–65°C. Specimens confined in an ovenlike space will bake and be useless. Specimens must remain on the dryer at the above temperature until they are crisp and brittle (but not baked or burned). *Never, never place mushroom specimens between newspaper sheets or in a plant press.*

Once dry, specimens must be kept dry or they will rehydrate and then become moldy and worthless. Removing freshly dried specimens directly from the dryer and putting them in a plastic bag large enough to accommodate them will help ensure that the specimens remain dry. In extremely humid regions, a small amount of desiccant can be added to safeguard against rehydration and mold growth. Delicate and fragile specimens can be dried in closed containers containing activated silica gel or other desiccant.

General Reminders

- Collect as many individuals as possible, including a range from young (immature) to old (mature or overmature). Four to five individuals will make an excellent unicate.
- Collect part of the substrate (leaf litter, wood, etc.) and note any phanerogamic association. Plucking while holding onto the stipe can destroy characters or leave part of the fungus behind.
- In preparing descriptions of macroscopic features, describe what you see on the basis of your experience. If in doubt, a sketch is extremely helpful. If at all possible, cut two or three sporocarps lengthwise in half before drying. If sporo-

carps are very fleshy, this should be done for all collected in order to promote drying.

Literature Cited

Beuchat, L. R. 1987. Food and beverage mycology. Van Nostrand Reinhold, New York.

Fidalgo, O. 1965. Conhecimento micológico dos índios brasileiros. Rickia **2:** 1–10.

——. 1968. Conhecimento micológico dos índios brasileiras. Revista Antropológica **15–16:** 27–34.

—— **& J. M. Hirata.** 1979. Etnomicologia Caiabi, Txicão e Txucarramãe. Rickia **8:** 1–5.

Findlay, W. P. K. 1982. Fungi: Folklore, fiction and fact. Mad River Press, Eureka, Calif.

Heim, R. & R. G. Wasson. 1959. Les champignons hallucinogènes du Mexique; Etudes ethnologiques, taxonomiques, biologiques, physiologiques et chimiques. Paris Muséum Nationale d'Histoire Naturelle, Paris.

—— & ——. 1965. The "mushroom madness" of the Kuma. Botanical Museum Leaflets **21(1).** Harvard University, Cambridge, Mass.

Parent, G. & D. Theon. 1978. Food value of edible mushrooms from Upper-Shabu region. Economic Botany **31:** 436–445.

Prance, G. T. 1984. The use of edible fungi by Amazonian Indians. 'In Ethnobotany in the Neotropics'. G. T. Prance & J. A. Kallunki, eds., Advances in Economic Botany **1:** 127–139. Bronx: The New York Botanical Garden.

Riedlinger, T. J. (ed.). 1990. The sacred mushroom seeker: Essays for R. Gordon Wasson. Dioscorides Press, Portland, Ore.

Wasson, R. G. 1969. Soma: Divine mushroom of immortality. Harcourt Brace & World, New York.

——. 1975. Mushrooms and Japanese culture. Transactions of the Asiatic Society of Japan **11,** ser. 3.

——. 1980. The wondrous mushroom: Mycolatry in Mesoamerica. McGraw Hill, New York.

7

Guidelines for Collecting Bryophytes

William R. Buck and Barbara M. Thiers

Institute of Systematic Botany,
The New York Botanical Garden

Introduction
Collecting Techniques
Preparing Specimens
Drying and Packing Specimens
Acknowledgments
Literature Cited

Introduction

Even though bryophytes are not a particularly important plant group from an ethnobotanical viewpoint, they are used in some areas as medicinals (Schultes & Raffauf, 1990), particularly in temperate regions (Johnston, 1987; Turner et al., 1983). Bryophytes are often perceived as taxonomically difficult, but most can be named with relative certainty. Furthermore, unlike higher plants, most bryophytes can be identified in their sterile condition.

Selected Guidelines for Ethnobotanical Research: A Field Manual, 143–146
Edited by Miguel N. Alexiades
© 1996 The New York Botanical Garden

Collecting Techniques

Bryophytes are among the easiest plants to collect. Only with extreme abuse do specimens lose their value. Although different workers have different field techniques, they are usually just variations on a theme. Because bryophytes have many growth forms, the specific collection procedure will depend on the plant at hand.

Use of a hand lens is essential for *critical* collecting of bryophytes. As in vascular plants, one can expect variation in the overall aspect of a bryophyte species, and this variation can appear to represent another taxon. Use of a hand lens will, in most cases, resolve any question, because leaf shape and other features of a size easily seen with a hand lens often remain constant. Also, some bryophytes are so small that to even be sure one has a bryophyte, a hand lens is necessary. If there is any doubt about whether one is collecting the same species repeatedly or collecting another, perhaps closely related, species, it is best to go ahead and make the collection.

Loose plants in tufts or mats can simply be picked up by hand, as can pendent forms. At times, however, plants are small or adhere tightly to the substrate. In such circumstances it is best to collect a small amount of the substrate with the plant to keep the plant from being lost or from breaking into many fragments. A knife or wood chisel is the best implement. Soil often stays around colonies of small, terrestrial plants. Indeed, it may be the plants that bind the soil, so soil should not be broken up, although some soil may be scraped from the bottom of the clod.

Small corticolous species are best collected with a shallow strip of bark so that slender stolons or stems that may be present will not be broken or lost. Clippers may be used to collect branch or twig sections with small bryophytes on them. As a general rule, for average-sized bryophytes, a piece the size of the palm of a hand is a good size for a single herbarium specimen. Additional material should be collected, when available, for sending to specialists, depositing in host country herbaria, and other functions (e.g., chemical or molecular analysis). There are small species, though, which, even when common, do not make large patches.

Some bryophytes occur only as scattered plants and should not be ignored just because of scanty, but well-developed, material.

General collectors should be careful when dividing material to send to specialists; bryophyte species, like those of higher plants, have distinctive aspects, but at times these may be too subtle for the untrained eye to discern.

Preparing Specimens

Bryophytes are best collected in individual paper bags, with a single collection in each bag. Although some collectors place large numbers of species into a single bag, individual species are often detected more easily in nature, and separation in the laboratory can be dirty and tedious. Plastic bags should never be used unless one intends to examine the specimens within a few days. Bryophytes, especially hepatics, are susceptible to becoming moldy if not allowed to dry or have air circulation. Also some mosses will continue to grow, but in a very odd, etiolated fashion, if left in plastic bags. In general, we have found that 2-lb, flat-bottomed, brown paper bags are the most convenient type. They are inexpensive and come bound in groups of 500 thus allowing easy transport and dispensability. One should purchase these before going to the tropics, since they are often either not available there or are of poor quality (e.g., with V-bottoms and glue that does not hold the bag together under tropical humidity). If paper bags are not available, packets can be folded from any available paper, or bryophytes can be put in a press with higher plants.

Although pressing does not alter critical morphological features, it does give the plant, especially mosses, a less-than-natural aspect when dry. In the case of leafy hepatics, the ideal method of preparation is to remove excess soil and other debris from the still moist (or rewetted) colony and place the plants between papers (or in a folded packet) in a plant press for about 24 hours (without heat). Tension on the press should be light.

Individual bags can also be used to record any desired field data. Although ecologists may be interested in complex habitat parameters, bryologists most often need only basic habitat information. Substrate is often recorded (e.g., tree trunk, moist soil, submerged on rock in stream), as is general habitat type (e.g., tropical rain forest, páramo, thorn scrub). Otherwise, label data should be as for other plants: locality (including latitude and longitude when possible), elevation, date, and collector (see Alexiades, Chapter 4, this volume). There is no need to give a descriptive account of the plant, as one often does for higher plants, since the whole plant is in the collection and almost never is altered on drying.

Drying and Packing Specimens

Although in very humid habitats it can be helpful to use a plant dryer, in most cases the bags can be left to air-dry. If in a vehicle, the paper bags can be put in a net or burlap bag and tied to the top of the car so that air is forced over them, facilitating drying. Very wet mosses, such as *Sphagnum,* or aquatic species, should be squeezed out, not wrung, to remove excess water. For both mosses and hepatics, air-drying is preferred, but limited exposure to heat in a mechanical dryer does not seriously diminish the value of the specimen.

In general, bryophytes are not excessively brittle, and dried plants do not need special care. An added advantage of individual paper bags is that they act as packing and insulation for the specimens in shipping. Usually just rolling up the bags and putting them in cardboard boxes is adequate protection. Some hepatics and *Sphagnum* (and a few other mosses) can be quite brittle on drying, so specimens should not be packed too tightly in the box, but they should be packed tightly enough to prevent shifting during shipping.

Literature Cited

Johnston, A. 1987. Plants and the Blackfoot. Lethbridge Historical Society, Lethbridge, Alberta.

Schultes, R. E. & R. F. Raffauf. 1990. The healing forest: Medicinal and toxic plants of the northwest Amazonia. Dioscorides Press, Portland, Ore.

Turner, N. J., J. Thomas, B. F. Carlson & R. T. Ogilvie. 1983. Ethnobotany of the Nitinaht Indians of Vancouver Island. Occasional Papers of the British Columbia Provincial Museum No. 24. British Columbia Provincial Museum, Victoria.

8

Collecting Bulk Specimens: Methods and Environmental Precautions

Douglas C. Daly
Institute of Systematic Botany,
The New York Botanical Garden
and
Hans T. Beck
Institute of Economic Botany,
The New York Botanical Garden

Selected Guidelines for Ethnobotanical Research: A Field Manual, 147–164
Edited by Miguel N. Alexiades
© 1996 The New York Botanical Garden

Introduction

The collection of bulk samples of plant material has become increasingly frequent as a component of ethnobotanical research. The traditional healer collects quantities of roots, bark, and other plant material to prepare phytotherapies. Scientists who make these collections are also interested in the medicinal properties of the plants, but from a different approach that usually involves studying the chemical compounds found in the samples. They may want simply to investigate the scientific basis for a widely used phytotherapy, or their ultimate goal may be the development of new scientific medicines based on purified compounds derived from the samples. Especially in the latter case, the bulk samples may be used to supply a series of studies including confirmation of bioactivity in in vitro assays, testing in additional screens, isolation of active principles, elucidation of chemical structures, toxicology studies, and clinical trials.

The goal of any scientific activity is to obtain valid, reproducible results. When plant material is being collected for chemical investigations, the methods employed must produce samples whose chemistry differs minimally from that of the living plant, and the methods might require collecting additional material. It must be possible to relocate the species, and, more important, the methods used in the initial collection must not irreparably damage the resource (i.e., the plant or plant population). In this context, the phrase "reproducible results" takes on a very literal meaning.

This chapter provides the field ethnobotanist with descriptions of preparations and procedures for the collection of bulk samples. We also provide perspectives on the environmental impact of this type of activity and approaches that will facilitate more advanced phases of pharmacognosy investigations. Thinking about some of these issues in advance can make the difference between back-breaking work with destructive and disappointing results, on the one hand, and back-breaking work that leaves some promise for future efforts on the other.

Logistics of Bulk Sample Collection
Obtaining Permits

Studies on medicinal plants must be carried out legally, ethically, and diplomatically. This type of research has become an extremely sensitive issue in recent years for a number of reasons; principal among them is the idea that for centuries Europe and later North America obtained commercially valuable medicines from plants collected in the developing countries, to whom the only "return" has been the sale of expensive drugs that are largely inaccessible to them economically and often irrelevant to their major health problems. More research groups in developing countries are participating more in the drug discovery process, but even as they increase their scientific capacity they often do not have the capital to follow through on development of scientific medicines and secure the necessary patents. In many instances, indigenous groups and other rural communities have had their lives and knowledge recorded and dissected by anthropologists or ethnobotanists, with nothing to show for it but cultural disruption. The fact that a monetary value can be calculated for a piece of genetic information in an organism—whether it codes for a medicine or confers disease resistance on a crop plant—has made developing countries more protective of their genetic resources (see Cunningham, Chapter 2, this volume).

The inconvenience of obtaining the appropriate permits is preferable to jeopardizing the entire research program and risking long-term censure. Even once you have secured the appropriate permits, be sure your purposes and activities are well known in the scientific community and the local communities in your study region.

Most countries require permits for collecting plant material of any kind, and some require special permits for collection of bulk samples for chemical analysis. Many require their own citizens to obtain permits. The regulations change frequently in some countries; be informed of the current guidelines well in advance of the planned research. In most countries, the requirements include a letter from the collaborating scientific institution, which is needed to obtain a permit from a government agency such as the forestry service, the department of agriculture, or a wildlife agency. Sometimes the permit can be obtained only in person in the capital city.

For material that will be exported for analysis, once the collections have been thoroughly dried and are ready for shipping, most countries require an export permit, a phytosanitary permit, and a certification from CITES (Convention on International Trade in Endangered Species of Wild Fauna and Flora). The destination country is likely to require the CITES permit as well as an import permit from the department of agriculture or a similar agency. The material will probably be inspected on arrival in the destination country, and if there is evidence of molding or of infestation with animals, the samples are likely to be impounded and perhaps incinerated. This is one of many reasons for careful selection, drying, storage, and shipping of the samples.

Timing of the Fieldwork

The optimal period for botanical fieldwork, especially when it involves the collection of bulk samples, is a function of several factors that depend to a large degree on precipitation patterns. Access to and from the study area is of primary importance, and, in many tropical regions, roads may be washed out by heavy rains in the wetter months, or low river levels may block access by water later in the dry season.

The height of the wet season must be avoided in more humid regions. All the time, effort, and funds invested in this type of fieldwork are lost if the bulk samples cannot be dried adequately—and kept dry—thus preventing mold until they can be stored under more controlled conditions. Moreover, visibility into the canopy is poor and tree climbing is more difficult and dangerous during rains in forest environments.

Phenology is another key consideration. The fieldwork should coincide with a peak in flowering or fruiting pattern in the vegetation, because "sterile" vouchers are often extremely difficult if not impossible to identify with certainty.

Virtually all habitats in temperate as well as tropical zones have a definable dry (or drier) season, and the available phenological studies indicate that the periods with the greatest number of taxa fertile (i.e., with flowers and/or fruits) often are the interfaces of the dry and wet seasons, especially the dry-to-wet season transition (Mori et al., 1982). Fortunately, this latter transition is also one of the best periods for road transitability, for collecting, and for drying.

Personnel

Ideally, at least four persons should carry out the actual fieldwork if significant numbers of bulk samples will be collected: one to record observations and codes; one to climb trees or otherwise gather voucher material (and then help with the bulk material); one to number and press vouchers; and at least one to collect, chop up, and bag the bulk samples. All of this work technically can be carried out by one person, but it should be kept in mind that the volume and weight of the bulk samples and vouchers accumulate rapidly. Under some circumstances it may be appropriate for local guides or informants to participate in the collection process.

Equipment

In addition to the tools normally used to collect voucher samples (see Alexiades, Chapter 4, this volume), it is important to include several other types of equipment for this work. Most of this equipment can be obtained from a good hardware or gardening supply center.

PROTECTIVE WEAR Gloves, long-sleeved shirts, long pants, and masks help protect the collector from various health hazards. Those preparing the samples should always wear gloves, because many plants have sap or hairs that irritate the skin. Some types of dermatitis are not immediate and therefore not obvious, appearing only after exposure to sunlight or after substances in the sap have attacked the skin over several hours (see also Labeling of Bulk Samples, below). Long-sleeved shirts and long pants also protect arms and legs from such irritants. A mask (or a bandana placed over the nose and mouth) helps reduce irritations from dust or plant hairs during collecting or drying procedures. Protective eyewear should be worn when operating equipment such as chainsaws or chippers.

FIRST-AID KIT There is always the danger of injuries due to cuts, splinters, and irritations. Bandages and antibacterial cream are musts. To remove splinters, thorns, or irritating hairs, a good pair of tweezers is essential; a magnifier will help you see them. Antihistamine or cortisone creams and topical anesthetics should also be included in case of allergic reactions.

DIGGING TOOLS A crowbar or nail puller or masonry hammer is an effective, lightweight tool to facilitate the excavation of roots of larger plants and collection of whole herbaceous plants. Shovels (including the collapsible army-issue types) and pickaxes are too bulky and heavy to be included in a standard list of field equipment, although they might be considered for larger-scale or longer-term projects.

CUTTING AND CHOPPING TOOLS The bulk material should be chopped as finely as possible to facilitate drying and to reduce the chance of internal molding. Several tools are useful for this purpose. Hand pruners work only on twigs or branchlets, so machetes (bush knife, cutlass) or lopping shears are needed for woodier material. Small, foldable pruning saws are also valuable for preparing woody material for bagging. Some collectors have been able to transport relatively lightweight manual composting mills for chopping up twigs, leaves, and other less woody tissues.

SCALES A spring balance is essential for weighing the samples before and after drying. Scales that can measure up to 5 kg in 50-g increments are optimal for bulk sample collecting (see Drying, below).

MARKERS Use felt-tipped markers with indelible ink to record identifying information on bulk sample bags as well as on the newsprint in which the voucher specimens are pressed. Waterproof ink is least likely to run, smudge, or fade. Ink that is insoluble in alcohol is preferable, because it can be used to label samples as well as vouchers that will be preserved in alcohol. Be sure to test.

any brand or color of marker you have not used before. In our experience, black is the most durable color. Felt-tipped markers can wear out quickly when used on cloth (see Labeling of Bulk Samples, below), so buy an ample supply (at least one for each week of fieldwork). Pencils or wax pencils (crayons) may also be used, but the results are not as legible.

BAGS If the material is not bagged, it will easily be lost or contaminated. Cotton is one of the best fabrics for the sample bags, because it "breathes" and is not easily ignited. Artificial fibers such as nylon or Tyvek can melt on contact with the screen above the heat source. The looser the weave of the fabric, the better the ventilation and the easier the drying; however, a weave that is too loose will result in escape of plant fragments and therefore loss of material or contamination from other samples. Therefore, onion bags and other plastic mesh bags are unsuitable.

Soil sample bags with a drawstring and a sewn-on label are ideal for this purpose (see Labeling of Bulk Samples, below); they can be obtained from forestry or scientific supply companies. If it is not possible to purchase bags before traveling, it is always possible to have them made in-country. One of the best fabrics is strong, densely woven, cotton mosquito netting, but even cotton pillowcases have served rather well. Drawstrings can be sewn on sample bags at little extra cost.

TARPAULIN Any type of tarpaulin can be used to help lay out the individual plant organs before bagging. A tarpaulin, however, while very useful, is optional and may be left behind if weight of equipment is a consideration. If plant material is spread out for preliminary sun-drying, one of the field personnel must remain with it all day in case of wind or rain, and care should be taken to ensure that each sample is correctly rebagged.

DRYERS See Drying, below.

THERMOMETER Excessive heat during drying can alter or destroy many of the chemical compounds in the bulk plant material. A long-stemmed thermometer, placed inside one of the sample bags, is useful for monitoring the temperature of the drying samples. The U.S. National Cancer Institute (NCI), which has been assaying plant extracts against various cancers (and now the HIV viruses) since 1960, asks its collectors not to dry bulk samples at temperatures exceeding 60°C (G. Cragg, pers. comm.). The equipment needed for drying is discussed below.

Collection of Bulk Samples

Recording Observations

The types of observations that are recorded in the field should facilitate not only identification of the plant but also return to the collecting site, or even return to the individual plant population or individual, in the event that additional material is needed for subsequent stages of a pharmacognosy investigation: chemical analysis, toxicology, clinical trials, and so on. It cannot be over-emphasized that all ethnobotany is founded on accurate identification of the plant species involved, and this depends on carefully prepared voucher specimens and carefully recorded observations. The more important types of observations to record for the vouchers are discussed by Alexiades (Chapter 4, this volume).

Other observations are related specifically to the bulk samples. For the traditional healer as much as for the natural products chemist, returning to a given plant or plant species ultimately means searching for the same active principle(s) the plant contained before. The composition and relative proportions of a plant organ's chemistry can be affected markedly by several factors (Ricker et al., 1994), and these should be recorded: the season, the time of day, whether the plant is in sun or shade, and the height on the plant from which the organs were collected (this is important for trees). Other factors, such as the habitat, the reproductive status of the plant, its size, and its health are observed as a matter of course for vouchers.

After the field observations for the voucher are noted in the field book, notes on the sampled plant organs are recorded as well. The number of bags and the fresh weight for each collected organ should be noted.

Minimizing Damage to the Resource

It is imperative that these collecting activities not be excessively prejudicial to the habitat or to the resources being sampled, for three reasons. First, such damage is wrong. Second, it is often illegal, and where it is not illegal it should be. Third, it is impractical: one or more samples will likely warrant further investigation, which inevitably involves re-collection on a larger scale; destructive initial collection is self-defeating because it may make re-collection impossible.

Herbs should be sampled only if an inconsequential fraction of the population is eliminated. The damage to perennial plants depends greatly on the organ(s) involved. For trees, careful removal of one or a few branch systems for fruits, leaves, twigs, or wood will usually have a negligible impact on the plant. Much more care is needed when bark or roots are needed. When these organs are involved, they should be removed from only one side of the tree. There is no way to eliminate the possibility of infection when bark is removed, although the risk may be reduced by using an antifungal paint (available from gardening and forestry supply companies) on wounds to trees. Still, a high proportion of tropical trees exude resins, latexes, or gums as a wound response, and these substances can effectively seal off a wound. In addition, many trees are capable of recovering from damage to the bark if they are not "girdled," that is, the cambium is not damaged over the entire circumference of the tree. This last consideration has been important in efforts to minimize damage to populations of the Pacific yew, until now the principal source of taxol (Daly, 1992).

The branch systems of many species of trees are dependent on the root systems directly below them; thus it is possible that the removal of roots from one side of a tree could kill that side, but killing one side is preferable to damaging some of the roots all around the tree, a procedure that would increase the risk of infection and undermine the structural support of the tree.

Selection and Collection of Plant Parts

After "safe sampling," local use is the ethnobotanist's most important criterion for sampling not only plant species but also plant parts. The basic premise is that empirical "testing" in traditional cultures over many years has led them to the species and organs with the desired physiological effects. For practical and scientific reasons, however, selection depends also on the habit of the plant, the plant population size, and the logistics of drying the material.

Medicinal plant preparations often include ingredients consisting of several plant organs from several species. For Western medicine-oriented studies, it is better in these cases to make separate samples of each component, because ultimately the investigators will want to know which bioactive compounds are produced in which species and in which plant organ(s). A truly rigorous approach would involve sampling the organs of a given species both separately and in the combination indicated by local use, in case there is a synergistic effect among compounds occurring in one ingredient but not another. Practical considerations may oblige the ethnobotanist to diverge somewhat from sampling strictly according to local use. For example, when local use indicates the aerial parts of an herb, it may be necessary to include the roots as well if the resource is limited. In other instances, it may be necessary to purposely omit extremely succulent organs from some combinations of plant parts if these samples would overload the drying system.

Unless an informant indicates use of damaged, infested, or diseased plants or plant parts, be sure to sample only healthy plants and plant organs. A pest or pathogen can alter the chemistry of the sample in two ways: first, the chemistry of the foreign organism will add "noise" to the results, and second, the chemistry of the plant itself can change in response to infection or attack.

Some considerations regarding plants with different habits are discussed below.

HERBS If the population or plants are large enough, roots and aerial portions can be collected as separate samples; otherwise, it is more common to collect the whole plant as a single sample. Some herbaceous species produce underground storage organs; if these are collected, samples should be clearly identified as taproots, bulbs, adventitious roots, tubers, corms, or rhizomes. Large storage organs of some aquatic herbs may be collected as separate samples. Especially large numbers of individuals of succulent herbs (see below) or aquatic plants must be collected to obtain even 500 g of dry weight; therefore it is crucial that the collector assess the size of the population before collecting. Certain large herbaceous species in families such as Araceae, Crassulaceae, Heliconiaceae, Strelitziaceae, Musaceae, and Rapateaceae can have very watery or mucilaginous stems and leaves; consequently, be sure that enough material is collected and that the bulk material is chopped finely and dried thoroughly so as to avoid fungal contamination.

CACTI, SUCCULENT PLANTS, AND SUCCULENT ORGANS Some of these plants contain large amounts of water-storage tissue, which may need to be excised and discarded in order to dry the sample. Cut succulent stems and leaves into many thin transverse sections for drying. It may take up to 2 weeks to thoroughly dry the stems of *Opuntia* or leaves of *Agave* species. In the case of succulent fruits, it may be difficult to completely dry the material; for example, a 13:1 weight reduction ratio was observed with the fruits of *Averrhoa carambola*.

EPIPHYTES For most epiphytes, the entire plant is collected. It is important to ensure that all individuals collected are of the same species. Some epiphytes have aerial roots, which may be collected as separate samples.

FERNS In general, ferns are sampled whole, although some terrestrial ferns have large rhizomes that can be collected as separate samples, and tree fern "trunks" can be sampled separately from leaves as long as the population can withstand sacrifice of the individual.

VINES Woody vines (lianas) often can be collected for their roots, woody stems, and leaves. Herbaceous vines are usually sampled whole, or the aerial parts may be sampled and the base and roots left intact if the population is sparse.

SHRUBS AND SMALL TREES One shrub or small tree may not provide enough material of roots or some other organs for an adequate sample, so if there appear to be enough individuals, several whole plants can be collected to make adequate samples of the roots, stems, and leaves or twigs.

TREES Trees can yield up to seven different plant organs for bulk samples: root bark, root wood, bark, wood, twigs, leaves, and fruits or seeds. Usually, however, only bark, wood, roots, and leaves or twigs are available in abundance. Note that the bark of many tropical montane trees and some lowland species is rather spongy with water and may lose 50% of its weight on drying.

Collection of Fresh Material

Fresh material is always preferable to dried material because the drying process causes the loss or modification of certain types of chemical compounds, notably monoterpenoids. If the field site is not far from a laboratory, or if express delivery services are accessible, it may be possible to deliver fresh material. This issue highlights the importance of building extraction laboratories in developing countries. Consultation with the laboratory doing the analyses can indicate whether the plant materials should be delivered in moist paper or in a container with ice or dry ice.

Transport of fresh material from one country to a laboratory in another is usually impractical because of legal and logistical difficulties. However, when local use specifies use of fresh plant material, the dried sample may be supplemented by an additional sample preserved in 70% ethanol in a watertight jar or bottle (e.g., Nalgene).

Labeling of Bulk Samples

Be sure to establish a standardized system of unique identifiers for the samples ahead of time. If bar codes will not be used, often it is sufficient to use the collector's initials and collection number, followed by a hyphen and a two-letter code for plant parts (see Table I for standard codes). Whatever system is used, a key to the codes should be written out in the field notebook. It may be advisable to use a plant-part code derived from the collector's na-

Table I. Suggested Plant Part Abbreviations for Bulk Samples

BD	bud	LF	leaf	SD	seed
BK	bark	PL	entire plant	ST	stems
CO	cone	PT	petiole	TU	tuber
EX	exudate	PX	entire plant without roots	TW	twigs
FL	flower	RB	bark of roots	UC	corm
FR	fruit	RH	rhizome	WD	wood
IF	inflorescence	RT	root	WR	wood of roots

tive language, if it is not English. As a check to avoid confusion in the records, we note the preliminary taxonomic identification on the label as well, and some collectors include a leaf of the plant in each sample.

Each bag of bulk sample material must be clearly and permanently labeled. If labels cannot be sewn on, you may use tags that are attached through the cloth: wired manila tags, baggage tags, or laundry tags can be tied on or attached with wire twists punched through the bag fabric. As a last resort, the information can be written directly on the bag, but oxidized plant saps can obscure this writing.

It is possible to purchase soil sample bags with a label made of no-tear, waterproof fabric sewn into the seam. The label information should be written with a waterproof marker. If many hundreds of samples will be collected, the manufacturer can be asked to design a customized label for the project. Any external label can be damaged or made illegible during handling, so a second label should be placed inside the bag.

If the plant sample is thorny or is known to be toxic or irritating to the skin, a hazard code should be used to advise those handling the samples. Such codes as TH for thorny, TX for toxic, and IRT for material irritating to the skin should be written conspicuously on the bag or bag label to warn sample bag handlers of potential hazards. You may want to use different-colored bags or attach colored flagging to this material as well. It is useful to pack all hazardous bulk samples into separate containers for shipment and to label that container appropriately.

Drying

Be sure not to collect more material than the drying system can handle. Although the vouchers can be preserved in ethanol and stored for several weeks before drying, in most instances the bulk samples should be dried immediately.

The type of dryer used for bulk samples will depend on the type(s) of energy available: solar, electricity, wood, or various liquid fuels. Before leaving for the field, verify what fuel(s) will be readily available, because this is the most important factor in determining your drying system.

Each energy source has its advantages and disadvantages. Solar power is free but difficult to harness and not always sufficient to combat relative humidity; moreover, ultraviolet radiation may affect the chemistry of the samples. Electrical power is safe and inexpensive but available only in towns or cities. Any flame is hazardous. Wood is often abundant, but smoke can affect chemistry. Liquid fuels are relatively efficient, inexpensive, and often easy to control, but they are very heavy.

SOLAR ENERGY   When possible, the material should be spread out in the sun for preliminary drying, as sun-drying considerably reduces the additional time needed over a supplemental heat source. Except in rather xeric climates, bulk samples will dry only partially in the sun. The material can be laid out on tarpaulins or in flat baskets. We are not able to comment on the solar ovens that are commercially available.

ELECTRICITY   When this is available, panels of light bulbs or electrical strip heaters are good choices for quickly and efficiently drying bulk samples. Strip heaters cost more to install than a light bulb system but are far more energy-efficient.

WOOD   Most wood-burning stoves or fireplaces are undesirable because most laboratories do not want samples contaminated with wood smoke. However, samples may be dried with the heat from a baking hearth or stove.

LIQUID FUELS   Portable cooking stoves or liquid fuel–powered space heaters are often used in the field for drying specimens because of their light weight, although the fuel itself is heavy. Such stoves and heaters can dry samples effectively, but caution must be exercised with the flames. Some cooking stoves burn only kerosene, some burn only white gas, and some burn any type of liquid fuel. Optimus and other pressure-type stoves are very fuel-efficient but need to be primed frequently and refilled several times over a 24-hour period.

   The diversity of drying systems is only as limited as one's ingenuity. Solutions have included baking ovens (including commercial ones, as they cool after business), wood-drying kilns, ceramic-firing kilns, and, in one very rustic situation, hanging bags from the rafters of a local family's home, high above the cooking hearth.

Bulk samples can be dried using frames similar to those used for drying vouchers (see Alexiades, Chapter 4, this volume). They must be sturdy enough to support the fresh weight of the samples. The samples should rest on a metal screen so that heat can pass through them but plant fragments cannot fall onto the heat source. The drying frames can be built with tiers of shelves made of wood-framed screens.

However the bulk samples are dried, the material should be chopped as finely as possible to hasten the drying process and prevent internal molding. During the drying process, each sample should be turned and mixed occasionally so that the material will dry uniformly. Samples should be dried in the bags; this method is slow, but it minimizes the danger of contamination or mixing with other samples. Otherwise, the material may be removed from the bags and placed directly on the screen, so long as they are adequately separated and labeled.

Bulk samples normally are dried at no more than 60°C. A long-stemmed thermometer should be used to monitor the temperature inside the sample bags, but if one is not available, a good rule of thumb is not to allow the plant material to become too hot to touch. Woody tissues lose about 10%–25% of their fresh weight when dried, whereas more succulent tissues such as leaves lose 35%–75% or more of their fresh weight; therefore, approximately 1 kg of woody organs and about 2 kg of wetter tissues are collected in order to obtain bulk samples consistently in excess of 0.5 kg dry weight (see also Cacti, Succulent Plants, and Succulent Organs, above).

The water content of the plant tissues also affects the amount of time required to dry the bulk samples. An efficient drying system can dry plant parts with low water content in 12–24 hours; extremely succulent tissues may require up to a week, especially if they are not finely chopped.

Large-Scale Re-collections

If an extract of a given species shows promise in bioassays or proves to contain interesting chemical compounds, it is inevitable that additional vouchered samples, possibly involving many kilograms dry weight, will be required for subsequent phases of the investigations. The ability to make accurate and effective large-scale re-collections depends greatly on the procedures used to make the initial samples and voucher collections (see Recording Observations, above). It helps if the original voucher collections are made on permanent study sites; in this case, the plants may be permanently tagged and thus easy to relocate.

When a large-scale re-collection is needed, there are several strategies for locating and recognizing the target taxon in the field. Vouchers can be consulted in an herbarium before returning to the field, and voucher fragments, photographs, or photocopies of vouchers can be taken along; this approach has worked very well in the past. It is wise to obtain taxonomic keys or descriptions from the literature when such references exist. Closely related species often occur together in lowland tropical forests, and consulting taxonomic literature and floras will help avoid confusion with similar-looking species.

In areas of current or recent ethnobotanical studies, on the return visit, the same informants can be reconsulted, and particular trees or individuals of target species can often be found via the common name(s) known to be associated with the species. Long-term residents in forest or other rural communities often have a remarkable capacity for returning to the same locality (and even the same plant) in a forest from which a collection was made. We have had repeated successes using this approach in our taxonomic work.

In unfamiliar locations, it is often possible to locate a species by providing a local informant with the local common name, which often can be found in ethnobotanical or forest-inventory literature or on existing herbarium collections. This is of course an imperfect approach: even within a given region, several names may be applied to the same species, and, conversely, one name may be applied to several species (Balée & Daly, 1990). You must confirm the identity of the species independently. Moreover, when the target species is relocated, new vouchers must be prepared in order to confirm the identity of the material being re-collected.

Although many lowland tropical species have broad distributions, results from quantitative forest inventories have shown that a species may be common in one area and rare in another, even nearby (e.g., Campbell et al., 1986). When a significant amount of material is needed for re-collection, it will be important to identify regions and (if possible) localities where relatively dense populations are likely to be found. There are two reasons. First, location of individuals of the species will be easier.

Second, large-scale collection will have less impact on a larger population. This kind of information can often be obtained by consulting the herbarium. The labels for many modern herbarium collections include information about the relative abundance of the plant in question; otherwise, species that are common in a certain region are usually represented by more numerous collections. Additional information may be obtained from the literature on forest inventories. Be sure to consult with local persons about the abundance or rarity of the species of interest.

The timing of large-scale re-collection is of critical importance. First, it is most important to collect during some part of the dry season, because the amount of material will necessarily involve initial sun-drying except in the rare instances when the material can be transported quickly to a drying facility. It will also be important that the taxon be fertile at the time of the second visit to ensure proper identification.

As noted above, selection of an appropriate locality for a bulk collection will be based partly on indications that the species in question occurs in relatively high densities there, so as to reduce the possible impact on the population. Before any collecting takes place in a given locality, a survey should be made of the population and an assessment made of how many individuals will necessarily be affected and what the damage will be to each. This will be especially important when small plants or large quantities are involved, and it is a crucial consideration for herbs, in which cases the entire plant is likely to be harvested. If the population might be damaged excessively by bulk collection, another locality should be sought. In some cases, it may be necessary to make larger-scale collections of a species at more than one locality.

Transport, Shipping, and Storage of Samples

For transport from the field, several bags can be sewn into a gunny sack with the collection number interval marked outside. If the processing area is in a town, the bags can be placed directly in strong cardboard boxes.

Although all voucher material except bark samples and very woody fruits can and should be packed in plastic bags, bulk samples must be shipped in "breathable" containers because many tissues are almost impossible to dry completely and enclosing them in plastic will promote molding.

When possible, the bundles of voucher specimens should be frozen for 48 hours to kill any pests; many large towns and most cities have businesses with commercial freezing facilities. The material should be inspected and placed in a cool, dry, insect-free storage area. Any sample showing more than a few spots of mold should be discarded, because significant superficial mold is usually an indication that the entire sample has been infected.

For international shipping, the samples should be sealed in heavy-duty cardboard boxes. It is advisable to attach copies of phytosanitary certificates, export permits, and import permits (or at least the import permit number for the institution in the destination country) on the outside of each box. If suitable heavy-duty cardboard boxes cannot be found, it is occasionally possible to have sturdy wooden boxes made by a local carpenter at low cost. Wood and bark samples and any other lignified tissues must be shipped in "breathable" containers, since it is virtually impossible to remove enough of the humidity to be able to pack them in plastic. The boxes are shipped air freight, air mail, or expedited air mail, or they can be taken as excess baggage. While surface mail is much less expensive than air freight, the risk of loss or damage is far greater.

Acknowledgments

We thank Miguel Alexiades and an anonymous reviewer for their helpful comments and corrections on the manuscript.

Literature Cited

Balée, W. & D. C. Daly. 1990. Ka'apor [Indian] resin classification. Advances in Economic Botany 8: 24–34.

Campbell, D. G., D. C. Daly, G. T. Prance & U. N. Maciel. 1986. Quantitative ecological inventory of terra firme and várzea tropical forest on the Río Xingu, Pará, Brazil. Brittonia 38: 369–393.

Daly, D. C. 1992. Tree of life. Audubon 94(2): 76–85.

Mori, S. A., G. Lisboa, & J. A. Kallunki. 1982. Fenologia de uma mata higrófila sul-baiana. Rev. Theobroma 12: 217–230.

Ricker, M. G. Veen, D. C. Daly, L. Witte, M. Sinta V., J. Chota I. & F.-C. Czygan. 1994. Alkaloid diversity in eleven species of *Ormosia* and in *Clathrotropis macrocarpa* (Leguminosae-Papilionoideae). Brittonia 46: 355–371.

III

Quantitative Methods in Ethnobotanical Fieldwork

Introduction

In the past years, a growing number of ethnobotanists have shown interest in quantifying plant–human interactions. Quantitative data can reveal information that complements qualitative observation, informing new research questions and helping in the process of making policy and resource management decisions. Furthermore, as Phillips remarks in Chapter 9, quantification can also stimulate the researchers to question their methodology and to deal with difficult questions that all too often are put aside. For example, quantification requires the consideration of sampling size and distribution, as well as the development of hypotheses that encourage data analysis within a theoretical framework. Many ethnobotanical studies have been limited to a purely descriptive level and have paid no attention to the distribution and variation of knowledge within a community.

Although quantification may be a clearly desirable goal in many cases, it is not without its problems and risks. Indeed, it is not purely due to academic laziness that many ethnobotanists have hesitated to embark upon quantitative initiatives. For one thing, although widely recognized as important, the question of how to deal with intracultural variation and sampling in ethnobotanical studies is not yet resolved. Another potential problem is that the "professional" appearance of numbers and statistics often belies the fact that quantification is only as good as the quality of the data on which it is based. Ultimately, the value of quantification will depend on how meaningful the data set is in relation to the questions asked. As outlined in Part I, it is often difficult to quantify cultural data, particularly if the researcher's categories do not correspond with those of informants. Ideally, ethnobotanists should have a means to cross-check the conclusions derived from their quantitative analysis, in order to ascertain that their results are not merely a construct of the method. Clearly, all these problems also exist in qualitative research, but they may be more prevalent and their consequences more ominous in quantitative approaches.

Phillips reviews several approaches used by ethnobotanists to quantify people's knowledge pertaining to plant use. Considerations pertaining to data analysis are important before commencing the study, as the type and assumptions of the analysis will frequently determine how the data need to be collected.

Although it is only recently that ethnobotanists have begun to grapple with the advantages and problems of quantification, anthropologists and human ecologists have long been quantifying the interactions between humans and their natural resources. As Zent notes in Chapter 10, human ecology and ethnobotany are kindred scientific endeavors, and ethnobotanists have much to gain from the experience and approaches of behaviorists and human ecologists. Zent presents a wide array of field techniques that can be used by ethnobotanists wishing to quantify people-plant interactions. Zent's contribution is particularly valuable as it complements the approaches listed by Phillips: whereas the latter deals with cultural knowledge, Zent presents techniques that utilize data derived from the direct observation of behavior. Zent underscores the importance of complementing data based on people's statements with the direct observation of behavior. Indeed, it is through this complementary juxtaposition of emic and etic approaches that many ethnobotanical interactions can be described and understood more fully.

In Chapter 11, Peters presents a series of ecological field techniques that ethnobotanists can use to address questions pertaining to the human use of plant resources and the impact of human–plant interactions on the resource base itself. Studies that incorporate a discussion of the spatial and temporal distribution of economically important plant resources and of the impact of human activity upon distribution not only will help advance the level and depth of ethnobotanical analysis but also will provide data of considerable value to development planners and policy makers.

The location of ethnobotany at the intersection of different fields is perhaps its greatest challenge and strength. Through the informed selection and use of quantitative anthropological and ecological field techniques, ethnobotanists can help resolve important questions concerning relationships between society and nature.

9

Some Quantitative Methods for Analyzing Ethnobotanical Knowledge

Oliver L. Phillips[1]
Missouri Botanical Garden

Introduction
Why Quantify?
  Data Collection
  Data Analysis
  Data Reliability
  Methodological Rigor
  Broader Scope
  Intellectual Property Rights
  Key Issues
Three General Approaches to Analyzing Quantitative Ethnobotanical Data
  Informant Consensus
  Subjective Allocation
  Uses Totaled
Choice of General Approach

[1] Current address: School of Geography, University of Leeds, Leeds LS2 9JT, U.K.

Selected Guidelines for Ethnobotanical Research: A Field Manual, 171–197
Edited by Miguel N. Alexiades
© 1996 The New York Botanical Garden

Introduction

Early this century, Kroeber suggested that ethnobotanical studies should become more quantitative (Kroeber, 1920), but his advice went largely unheeded by ethnobotanists for more than 60 years. The impact of quantitative techniques on ethnobotany is finally being felt, and it has been a key factor in reinvigorating a field that has been scientifically marginalized for too long. Since the mid-1980s, numerous papers have incorporated quantitative analysis to address a wide range of issues (see Table I).

For the purposes of this review, I define quantitative ethnobotany as "the direct application of quantitative techniques to the analysis of contemporary plant use data." I will address one specific aspect of quantitative ethnobotany: the analysis of people's knowledge of plant use. This chapter reviews techniques that were devised to answer questions such as How significant is use x, species y, or plant community z, to people? The techniques mostly take a "plant-centric" approach and are most appropriate for ethnobotanical research primarily oriented toward botanical, conservation, or pharmaceutical, rather than strictly anthropological, goals. Quantitative analyses in cognitive ethnobiology and of purely anthropological or biological phenomena are excluded, although quantitative techniques are becoming increasingly important in all branches of ethnobiology (Berlin, 1992; Romney et al., 1986). Methods to collect and analyze quantitative information primarily based on behavioral or ecological data are covered in Chapters 10 (Zent) and 11 (Peters). Finally, analyses of the commercial significance of plant uses are also excluded from this review since they are usually considered part of "economic botany."

The context in which ethnobiological data are collected clearly

Table I. Selected investigations in quantitative ethnobotany

Method	Reference	Focus of research
Informant consensus: The relative importance of each use is calculated directly from the degree of consensus in informants' responses.	Adu-Tutu et al. 1979; Elvin-Lewis et al. 1980	Age, sex, ethnic group, as determinants of chewing-stick preference
	Angels Bonet et al., 1992	Frequency of medicinal plant use
	Friedman et al. 1986	Identify ethnomedically important species
	Johns & Kimanani, 1991	Test a model of origin of traditional medicine
	Johns et al. 1990, 1994	Identify ethnomedically important species
	Joly et al. 1987	Identify most important uses
	Kainer & Duryea, 1992	Identify most important uses and species
	Perez Salicrup, 1992	Compare forest and nonforest species for commerce and subsistence use
	Phillips & Gentry, 1993a	Evaluate usefulness of tree and liana species and families
	Phillips & Gentry, 1993b	Ecology, phylogeny, physiognomy, growth rate, and habit, as correlates of plant usefulness; knowledge as correlate of age
	Phillips et al., 1994	Compare usefulness of forest types; compare performance of quantitative ethnobotanical methods
	Trotter & Logan 1986	Identify ethnomedically important species
Subjective allocation: The relative importance of each use is subjectively assigned by the researcher.	Berlin et al. 1966, 1974	Cultural significance and lexical differentiation
	Lee 1979	Characterize plant resources
	Pinedo-Vásquez et al., 1990	Percent useful species per area of forest; evaluate destructive vs. nondestructive uses
	Prance et al. 1987	Percent useful species per

Table I. Selected investigations in quantitative ethnobotany (*continued*)

Method	Reference	Focus of research
		area of land; most important species and families; compare knowledge of different ethnic groups
	Stoffle et al., 1990	Compare plant importance with "land-use area" importance, to set conservation priorities
	Turner 1974	Characterize plant resources
	Turner 1988	Cultural significance of plants
Uses totaled: No attempt is made to quantify the relative importance of each use. The numbers of uses (or "activities") are simply totaled, by category of plant use, plant taxon, or vegetation type.	Anderson 1990, 1991	Percent useful trees per area of managed and unmanaged forest
	Anderson & Posey 1989	Percent useful plants in managed scrub
	Balée 1986	Percent useful trees per area of forest
	Balée & Gely 1989	Percent useful plants in managed succession
	Bennett, 1992	Percent useful trees per area of forest
	Boom 1985, 1989, 1990	Percent useful trees per area of forest
	But et al. 1980	Percent medicinal species vs. plant family size
	Bye, 1995	Compare regions for plant use; compare importance of different kinds of use
	Carneiro 1978	Percent useful trees per area of forest
	Kapur et al., 1992	Identify most important plant families for traditional medicine
	Moerman 1979	Selectivity in medicinal plant choice
	Moerman, 1991	Medicinal plant use as a cor-

Method	Reference	Focus of research
		relate of plant family, subclass, growth habit, and life form
	Paz y Miño et al., 1991	Liana use as a correlate of physiognomic and ecological factors
	Salick, 1992	Compare forest types for plant use
	Toledo et al., 1992	Compare plant products by indigenous group and habitat
	Unruh & Alcorn 1988	Percent useful plants in managed succession

can affect the research conclusions, whether the analysis is quantitative, qualitative, or both. Interviewing is an essential skill for botanists conducting ethnobotanical or ethnomedical field surveys in species-rich areas and for conservation planners needing useful information about the subsistence use of species or ecological communities. Such information on plant uses can be gathered much more rapidly through interviews than by direct observation alone. Interpreting interview data, however, is not always straightforward. The researcher needs to understand in advance how the interviewing process itself might affect the "results" and must take steps to minimize or control for these dangers (Alexiades, Chapter 3, this volume). A fundamental issue in any ethnobotanical study is the degree to which the data gathered reflect the informants' perceptions of their environment. So, although this chapter presents apparently sophisticated analytical procedures, the reader should beware of applying them uncritically. The old computer adage of "garbage in, garbage out" can be equally applicable to ethnobotanical analysis if the raw data are seriously flawed!

Why Quantify?

Quantification should not be seen as an end in itself but rather as a means to address particular ethnobiological questions. Perhaps the greatest danger with this approach is that by working with numbers and statistics we risk forgetting that their validity ultimately depends on the quality of the data used to generate them. Unconsciously or consciously, one could use statistical analyses to mask major problems with the original interview data. Furthermore, at some point in any critical analysis of ethnobiological data we are usually obliged to construct subjective categories (for example, the categories "food plants" and "medicinal plants") that do not necessarily correspond to those of the informants, and the danger is that we may come to construe our categories as theirs. However, quantitative analysis does not necessarily require any more subjective categorization than does qualitative analysis. We should also remember that arguments about whether we are really measuring the "right thing" are hardly unique to the case of quantification in ethnobiology. This controversy largely mirrors the history of the development of most natural and social sciences, most of which have gone on to embrace quantification and statistical methods. For example, in cultural and ecological anthropology there is now a quantitative tradition, and most of the earliest quantitative ethnobotanical studies had a distinctly anthropological perspective (Berlin et al., 1966, 1974). Other fields with direct relevance to ethnobotany—such as systematics and ecology—have been repeatedly revolutionized by the impacts of quantification and hypothesis testing. Perhaps one obstacle to the earlier application of quantitative methods to ethnobotany was a widespread perception that to quantify implies a reductive, even clinical way to study the interaction between people and plants. Yet, quantification and hypothesis testing are by no means necessarily reductionist. As well as helping to dissect the often complex nature of plant–human interaction, they can also be used to build and test integrative models (for example, concerning the origins of plant uses; Johns & Kimanani, 1991; Moerman, n.d.) or to compare the cultural significance of whole ecosystems.

Ultimately, quantitative ethnobotanical techniques are complementary to the more traditional forms of ethnobotanical inventory; they are not alternatives to them. Although quantitative techniques allow us to analyze patterns of plant use knowledge, in a full ethnological study they cannot replace the need for careful qualitative description of indigenous knowledge. Moreover, quantitative analysis cannot eliminate any inherent biases in the data collection process, although statistical analyses can help in assessing the potential influence of such biases. And, however refined the analytical technique, the results always need to be interpreted in reference to the data collection procedure. Ideally, if time permits, the solution to these problems will lie with the investigator first developing a detailed ethnography. Only once a broad understanding of the cultural context of plant use has been achieved should the investigator classify plant utilization in a form suitable for quantitative analysis. Where time constraints preclude in-depth ethnographical study, ethnobotanists need to be especially aware that any quantitative analysis will be partly based on subjective constructs.

Although we need to be aware of the dangers involved in quantification, there are greater dangers involved in *failing* to consider new quantitative methods. I suggest that the benefits of adopting quantitative approaches to analyzing ethnobiological data usually far outweigh the risks. There are at least three clearly identifiable primary benefits, involving the collection, the analysis, and the reliability of ethnobiological data. Together their effect is to improve the methodological rigor and broaden the scope of ethnobotanical studies (two additional benefits). A sixth benefit, related to the protection of intellectual property rights of informants, is also sometimes realized. The six benefits are discussed in more detail here.

Data Collection

Deciding from the outset that the information will be analyzed quantitatively encourages the ethnobotanist to carefully consider how they should be collected (Johns et al., 1990), because quantitative analysis demands that data be collected in as explicit, repeatable, and systematic a way as possible. Close consideration of data analysis methods may forewarn the researcher of informant sampling issues that need to be considered, which might not otherwise have been so obvious. For example, if the aim is to describe the distribution of knowledge within and among households, then considering the statistical issues involved will compel the researcher to reflect on possible complicating factors and to decide, for example, what will be the minimum number of informants and households needed to properly address the research questions.

Data Analysis

Numerical data can be analyzed statistically, and confidence intervals and probability levels can be assigned to the values obtained. Statistical analysis has become an integral part of the core scientific process that involves formulating and testing falsifiable hypotheses (Popper, 1963).

Data Reliability

Statistical analyses may also allow the investigator to check the reliability of the data collected. Reliability has been a serious issue in ethnobotany because results from working with just a few informants were often extrapolated uncritically to whole communities or cultures. Again, simple descriptive statistics giving information on the confidence one has in the results are the norm in the rest of science. With quantitative data analysis, confidence intervals can easily be applied to ethnobotanical data. For example, in one study (Phillips & Gentry, 1993b), we calculated confidence intervals for a species' use value estimate to make explicit the relationship between the number of informants interviewed and statistical confidence in the estimated use value.

Methodological Rigor

As a consequence of the first three benefits, ethnobotany will benefit from enhanced scientific status as more exacting approaches (e.g., quantitative, statistical, model-building, and hypothesis-testing) are applied. Greater scientific status should help ethnobotanists attract the increased funding urgently needed to investigate cultural and biological diversity.

Broader Scope

Quantification also enables ethnobotanists to broaden the scope of the field. For example, specific hypotheses concerning the relationships between knowledge of plant use and critical variables such as informant age, sex, or profession, can be investigated statistically (Phillips & Gentry, 1993a). Other issues are also being addressed quantitatively (see Table I). For example, plant use data have recently been used as the basis for ranking the importance of cultural resources threatened by development (Stoffle et al., 1990), to assess the efficacy of particular remedies for particular diseases (Johns et al., 1990, 1994), to test a model of the origins of medicinal plant use (Johns & Kimanani, 1991), and to test several hypotheses about factors that determine a plant's usefulness (Phillips & Gentry, 1993b).

Intellectual Property Rights

Finally, one outcome of much traditional ethnobotanical research is a listing of plant species and their precise uses. While this exercise often has intrinsic merit, it may put the intellectual property of informants squarely in the public domain, with perhaps unintended consequences for the exploitation of that knowledge. Quantitative analytical approaches can make this kind of one-to-one matching of species to use redundant (i.e., of little additional academic value). Quantitative approaches might be less likely than the listing approach to compromise the intellectual property rights of informants. Of course, this is not necessarily the case (see below), and in any event it does not reduce the ethical obligations researchers have to their informants and host cultures.

Key Issues

Two particular issues have motivated many studies in quantitative ethnobotany, as defined here. Both have been explored before in qualitative terms, but quantification has provided new insights. The first issue, the degree to which traditional people use their total forest environment, was addressed by Carneiro (1978) and more rigorously since by other researchers (e.g., Anderson, 1990; Balée, 1986; Boom, 1985; Phillips et al., 1994; Pinedo-Vásquez et al., 1990; Prance et al., 1987). Often these studies are embedded in an explicitly conservation–oriented research agenda and explore such questions as the dependence of people on different vegetation types or the ability of people to create and manage productive, species-rich ecological communities. By combining quantitative ethnobotanical and ecological data, ethnobotanists have highlighted the importance of tropical forests to traditional peoples and, conversely, the profound effects that indigenous people may have on "wild" vegetation (Anderson & Posey, 1989; Balée & Gely, 1989; Boom, 1985). This focus on the strong relationship between cultural and biological diversity has provided greater incentive for conservationists and indigenous people's rights or development groups to work together for common goals.

Another common concern of quantitative ethnobotanical studies is assessing the relative cultural significance of many medicinal plants, often with the specific goal of identifying biomedically efficacious species—a goal with potentially important financial and intellectual property ramifications. Different approaches to answering questions of this kind have been developed (Adu-Tutu et al., 1979; Friedman et al., 1986; Johns et al., 1990; Moerman, 1991; Phillips & Gentry, 1993a,b; Trotter & Logan, 1986).

Three General Approaches to Analyzing Quantitative Ethnobotanical Data

Either explicitly or implicitly, all researchers assume that there is a direct relationship between what people say is important (i.e., the interview data) and what is culturally significant. Although we might expect this assumption to be broadly valid, I know of no ethnobotanical studies that statistically describe the relationship between quantitative plant use information derived from individual interviews and that derived from behavioral data. Such comparative methodological research is needed to test this fundamental assumption of the interview method.

The quantitative approaches covered here for analyzing informant knowledge can be grouped under three categories (Phillips & Gentry, 1993b): informant consensus, subjective allocation, and uses totaled. Here, descriptions of these categories are followed by suggestions for choosing an appropriate method and a brief description of each published method.

Informant Consensus

In the informant consensus method, the relative importance of each use is calculated directly from the degree of consensus in informants' responses. The importance of different plants or uses is assessed by the proportion of informants who independently report knowledge of a given use or who claim to have used a plant in a specific way (Adu-Tutu et al., 1979; Angels Bonet et al., 1992; Elvin-Lewis et al., 1980; Friedman et al., 1986; Johns et al., 1990, 1994; Johns & Kimanani, 1991; Joly et al., 1987; Kainer & Duryea, 1992; Perez Salicrup, 1992; Phillips & Gentry, 1993a,b; Phillips et al., 1994; Trotter & Logan, 1986). This approach is amenable to statistical testing and is relatively objective. However, it is also time-consuming because individual informants or households must be interviewed separately. Data suitable for analysis by informant consensus are typically collected in independent interviews of individual informants. All interviews are usually conducted under as comparable conditions as possible, and individual informants may be reinterviewed during the course of the study.

Subjective Allocation

In the subjective allocation method, the relative importance of each use is subjectively assigned by the researcher. The importance of different plants or uses is estimated by the researcher on the basis of his or her assessment of the cultural significance of each plant or use (Berlin et al., 1966, 1974; Lee, 1979; Pinedo-Vásquez et al., 1990; Prance et al., 1987; Stoffle et al., 1990; Turner, 1974, 1988). This is a quicker approach than informant consensus for evaluating the cultural significance of plants. Yet, compared with informant consensus, the results are more subjective and less amenable to statistical analysis. Data for this kind of analysis can be collected by one or more interview techniques or by direct observation or by both.

Uses Totaled

In the uses totaled method, no attempt is made to quantify the relative importance of each use. The number of uses (or "activities") are simply totaled, by category of plant use, plant taxon, or vegetation type. Not surprisingly this has been the most popular approach, since it is the fastest and most straightforward way to quantify ethnobotanical data (Anderson, 1990, 1991; Anderson & Posey, 1989; Balée, 1986; Balée & Gely, 1989; Bennett, 1992; Boom, 1985, 1989, 1990; Bye, 1995; Kapur et al., 1992; Moerman, 1979, 1991; Paz y Miño et al., 1991; Salick, 1992; Toledo et al., 1992; Unruh & Alcorn, 1988). But it has two principal disadvantages. First, minor uses are treated as equivalent to even the most important of uses. Second, the total numbers of uses recorded may be more a function of research effort than of the relative significance of each use, plant, or vegetation type. Like those for subjective allocation, data for this kind of analysis are often collected with one or more interview techniques, and sometimes by direct observation.

Choice of General Approach

I suggest that the investigator first decide which of the three general approaches is most appropriate given the aims and the practical constraints of the research. Once this basic decision has been made, the choice of which exact technique to use (see A Brief Description of Each Published Technique, below) should be easier to make. Some important aspects of each of the three approaches are described in Table II. All the methods require a certain amount of subjective judgment about what categories are appropriate for analysis, so none of the methods should be applied uncritically to ethnobotanical data.

Nevertheless, using informant consensus as a measure of cultural significance has a number of advantages over procedures that subjectively assign importance values a posteriori or simply add up numbers of uses. (1) Informant consensus use values reflect the importance of each use or species to each informant, since, in most situations, it is reasonable to assume that the greater the salience of a given use or plant, the more likely it will be mentioned. "Wrong," or even deliberately misleading, answers will tend not to have a substantial influence on the results, because, by definition, this methodology assigns greatest value to those uses confirmed in multiple interviews with different people. (2) Informant consensus makes efficient use of all available information, because every interview contributes directly to the use value calculation even if it provides only "negative" data by which a plant is claimed to have no use or is not recognized. (3) The distribution of use values generated is continuous, not discrete. This continuity lends greater power to statistical tests because continuous data contain more information than equivalent discrete data. (4) Use values are unbiased by the intensity of research, because results per use or per plant are expressed in proportional terms compared with the number of interviews. (5) Statistical confidence intervals can be calculated for each use value. Similarly, a quantitative description of the relationship between sample size and the accuracy of the use value estimate can be derived (see Phillips & Gentry, 1993b: fig. 3). Thus, statistical criteria for determining an adequate sample size can be made explicit. (6) Where the sample size of informants may be inadequate, the confidence intervals for each species' estimated use value can be narrowed by resampling techniques such as bootstrapping or jack-knifing (Magurran, 1988). (7) Direct subsistence and commercial uses (equivalent to consumptive use value and productive use value, respectively, sensu McNeely et al., 1990) can be evaluated simultaneously by the same measure, allowing comparison of the importance of each form of production. By their very nature, subsistence economies are invisible under financial accounting procedures. Therefore, if development projects are to be properly evaluated, we need techniques to simultaneously measure subsistence and commercial value. Informant consensus provides one way to do this, by allowing the researcher to express commercial value in terms of a *subsistence* "use value" framework (Phillips et al., 1994).[2] (8) The method by which plant importance is assessed is completely transparent, and, in theory, replicable (i.e., two studies with exactly the same interview data would yield exactly the same results). The method is replicable because the estimates of relative importance are explicitly and numerically derived from interview data rather than assigned a posteriori (subjective allocation) or nonexistent (uses totaled). The greater rigor of the method gives us greater confidence in the results of hypothesis testing.

[2] Several discussions and critiques are available of the many techniques that economists have developed to estimate subsistence use value in commercial terms (Brown, 1994; Chopra, 1993; Godoy et al., 1993; Pearce, 1993).

(text continued on page 186)

Table II. A comparison of approaches used to estimate knowledge of plant use (from Phillips & Gentry, 1993a)

	Informant consensus	Subjective allocation	Uses totaled
Methods explicit?	Yes	No (relative importance is assigned *a posteriori* by researchers)	Rarely (research effort is rarely quantified)
Objectivity of use values generated	Relatively objective	Relatively subjective	Relatively subjective
Distribution of use values generated	Continuous	Usually discrete	Discrete
Negative data useable in analysis?[1]	Yes (negative data is used directly to calculate use values)	Yes (negative data may help differentiate importance of uses)	No
Statistical comparisons possible between uses?	Yes (score from each species or informant represents one observation)	No	No
Statistical comparisons possible between species?	Yes (score from each informant represents one observation)	No	No

Statistical comparisons possible between informants?	Yes (knowledge of well-studied species can be compared)	Yes, but less precise than A; no such study known	Yes, but less precise than A; no such study known
Statistical comparisons possible between ethnic groups, plant communities?	Yes, and such comparisons could be valid even with different researchers	Yes, but such comparisons are less valid with different researchers	Yes, but interpretation is difficult because no two studies involve an equal level of research effort
Valid comparisons possible between subsistence and commercial uses?	Yes, the relative importance of each can be directly determined from the informants' responses	Unlikely (researcher value-judgements may obscure the comparison)	Unlikely (one commercial use is unlikely to be of equivalent importance to one subsistence use)
Speed of data collection and analysis	Most time-consuming	Less time-consuming	Potentially the least time-consuming

[1] "Negative data": defined as informant either not recognising plant, or recognizing it but knowing no uses.

In spite of these advantages, if time is a severe constraint or a large number of taxa are to be evaluated, the simplest approach (uses totaled) can be the most appropriate. But if the research will be long-term and in-depth and if there are opportunities to work with several knowledgeable informants, it is certainly worth collecting the data in a form that can be analyzed by informant consensus. And if two or more investigators are recording data, or if the aim is to make cross-cultural comparisons with data gathered by other researchers, then collecting information by a standardized format and analyzing it by informant consensus will increase the validity of the conclusions. Alternatively, if the researcher is already familiar with the culture, then subjective allocation may be a more time-effective approach. Indigenous ethnobotanists, in particular, would have a lifetime of familiarity with the culture, so their value judgments on the relative importance of different plants would clearly carry weight.

A Brief Description of Each Published Technique

Once the choice of general approach has been made, a variety of individual quantitative analytical techniques are available. Here, for each of the three general approaches, I briefly describe the essence of each distinct technique. Since it is not possible to do full justice to each specific technique that has been used to quantify plant use data, I urge the reader to consult the primary sources to fully understand the context and nuances of each. At the start of some sections, the papers that used the same analytical technique are listed together in brackets. Those works and others using the same general approach, as well as the issues that the authors were investigating are described in Table I.

One more note: some of these methods may appear complicated at first, but in practice most are quite simple to compute. By substituting some numbers, you can quite easily dispel the mathematical mystique of complex-looking equations.

Informant Consensus

The relative importance of each use is calculated directly from the degree of consensus in informants' responses.

[Adu-Tutu et al., 1979; Elvin-Lewis et al., 1980] Informants were interviewed with the aid of a highly structured questionnaire as to their preferred species for a particular medicinal (chewing stick) use. The relative importance of each species was evaluated by the proportion of respondents who cited it. Supplemental questions allowed the researchers to distinguish statistically between different factors influencing the informants' choice of species. This approach is unusual among quantitative ethnobotanical studies in that the authors analyzed data gathered by use of a rigidly structured questionnaire, designed to elicit information in predetermined categories. Although questionnaires are well suited to addressing particular focused research questions, they might prove somewhat limiting if the aim of the research is to characterize all the group's knowledge (see Alexiades, Chapter 3, this volume).

Friedman et al. (1986) used a technique designed to highlight species that have healing potential for specific major purposes and merit further biomedical research. Interviews took two forms. In the first, open-ended interviews were used to encourage each informant to separately volunteer information on medicinal plants. In the second, more structured, interview format, a list of 50 commonly used species was read out and the informant was asked to provide information on any familiar species in the list. The authors scored each plant's popularity among the sampled population according to the number of informants who cited it. For each plant, they also calculated an index (called fidelity level, FL) designed to quantify the importance of the species for a particular given purpose. Each plant species' FL was simply calculated as the ratio between the number of informants who suggested the use of a species for the same major purpose (I_p) and the total number of informants who mentioned the plant for any use (I_u). Hence:

$$FL = \frac{I_p}{I_u} \times 100\%$$

For each plant, a derived measure (rank order priority, ROP) was also calculated from the combination of its FL and its overall relative popularity. This technique could easily be modified to quantify differences in a given species' importance for particular purposes to different sexes, age groups, or to other ethnic groups, by calculating *FLs* separately for each group.

Trotter and Logan (1986) used numerical techniques to help test the hypothesis that there should be greater informant consensus about remedies that are pharmacologically effective than about those that are not. Informant consensus for each remedy was evaluated in two ways: first, by the relative frequency with which each *remedy* was mentioned in interviews and, second, by calculating an informant agreement ratio (*IAR*) for each *ailment* by the following method:

$$IAR = \frac{n_a - n_{ra}}{n_a - 1}$$

where: n_a is the total number of cited cases of the ailment and n_{ra} is the total number of different remedies for that ailment. The maximum *IAR* is 1, when there is complete agreement among informants about the particular remedy for a particular ailment (i.e., when $n_{ra} = 1$, and $n_a \geq 2$). The minimum score is 0, when as many different remedies are cited as there are reports of the ailment (i.e., when $n_{ra} = n_a$). Trotter and Logan's techniques measure informant consensus about favored treatments for specific ailments. They are relatively easy to apply, and preliminary results of the authors' experiments suggest that they may also be useful tools for helping to identify pharmacologically active remedies.

[Johns et al., 1990, 1994; Johns & Kimanani, 1991] Johns et al. assumed that common species or diseases are more likely to be mentioned in interviews than are rare species or diseases. They then applied a log-linear model to their data to quantify informant consensus about the best remedies for particular diseases, independently of the confounding effects of plant and disease frequency:

$$\log\ (n_{ij}) = \mu + \alpha_i + \beta_j + \tau_{ij} + \epsilon_{ij}$$

where

$n_{ij} =$ the total number of people who independently said that plant species i is used to treat disease j

μ = an overall effect
α_i = the main effect due to plant i
β_j = the main effect due to disease j
τ_{ij} = the interaction effects, indicating the potential for plant i as a cure for disease j
ϵ_{ij} = random variation in cell (i,j)

In the model, τ_{ij} is a quantitative measure of the degree of confirmation of any particular remedy. The mathematical notation used in the log-linear model is explained in Bishop et al., 1984.

[Kainer & Duryea, 1992; Perez Salicrup, 1992] These authors estimated the importance of each individual use as the ratio of the number of informants who mentioned a given species' use to the total number of informants interviewed about that species. These were the first authors to apply an informant consensus approach to evaluating a species' *overall* usefulness rather than just its medicinal importance. The technique's clear advantage is its simplicity.

[Phillips & Gentry, 1993a,b; Phillips et al., 1994] We used a two-tiered approach to quantify a species' importance. First, the use value of each species for each informant (UV_{is}) is estimated as:

$$UV_{is} = \frac{\Sigma U_{is}}{n_{is}}$$

where U_{is} is the number of uses mentioned by informant i for species s in each interview, and n_{is} is the number of interviews with informant i for species s. Thus, the use value of each species for each informant is defined as "the ratio of the number of uses mentioned in each interview, totaled for all interviews, to the number of interviews for that species." Our estimate of the overall use value for each species (UV_s) is then:

$$UV_s = \frac{\Sigma_i UV_{is}}{n_s}$$

where n_s is the number of informants interviewed for each species. So, each species' use value is simply "the average of each informant's use value for that species."

The cultural importance of plant families can also be compared by estimating family use value (*FUV*) by:

$$FUV = \frac{\Sigma\,UV_s}{n_f}$$

where n_f is the number of species in the family. The *FUV* is thus "the average use value of each species in the family." (Prance et al. [1987] and Phillips and Gentry [1993a] discuss other possible ways to estimate the cultural significance of plant families.) These techniques were developed to analyze information gained from reinterviewing the same informant at intervals of several months or years. Long intervals between interviews with the same informant help ensure that the results of the interviews are broadly independent. Thus, this approach is most suitable for a long-term project in which the cultural significance of many species is to be assessed.

Phillips and Gentry (1993b) compared the relative knowledge of different informants by calculating a standardized relative use value (*RUV*) for each informant:

$$RUV = \frac{\Sigma\,\dfrac{UV_{is}}{UV_s}}{n_{is}}$$

where n_{is} is the number of folk-species with data from three or more informants, calculated for each informant. Recall that the use value of each species for each informant is UV_{is}, and the overall use value for each species (for all informants combined) is UV_s. So, by averaging UV_{is}/UV_s for all species, the index evaluates the overall plant use knowledge of each informant compared with that of other informants. The same principle could be applied to generate similar indices from other kinds of informant consensus data.

Subjective Allocation

The relative importance of each use is subjectively assigned by the researcher.

[Berlin et al., 1966, 1974; Lee, 1979; Turner, 1974] These studies share a semiquantitative element and the basic assumption that the more highly managed a plant species is, the more culturally significant it must be. Thus, Berlin et al. (1966) and Turner (1974) classified species' cultural significance as low, moderate, or high; Berlin et al. (1974) defined four levels of cultural significance (cultivated, protected, wild but useful, and culturally insignificant), and Lee (1979) defined six. To varying degrees, the categories used by these authors were based on informants' concepts. These estimates of management intensity are appropriate for research designed to include the ethnobotany of the whole spectrum of human management of plants—i.e., from crops to unmanaged wild plants.

[Pinedo-Vásquez et al., 1990; Prance et al., 1987] The uses of forest plants were classified as minor or major; each minor use was given a score of 0.5, and each major use a score of 1.0. Use scores were then summed for each species to calculate each species' use value. Within each plant family, species' use values were averaged to calculate a family use value. Similarly, the relative importance of subjectively assigned use categories (edible, construction, technology, remedy, commerce, and other) were calculated for each family and for four indigenous groups. The Prance et al. paper is a benchmark study in quantitative ethnobotany, being the first to systematically address the question, How important is the forest to indigenous people? One difficulty with this technique, as used, is that it does not allow for more than one use for each species within each category. The various quantitative approaches to ethnoecological forest inventories are compared in Phillips et al., 1994.

Turner (1988) developed an index of cultural significance for each plant species. The index is a composite of a variety of plant use attributes as perceived by the researcher (use importance, intensity, and exclusivity). Each species' index of cultural significance (*ICS*) is calculated as:

$$ICS = \Sigma \, (qie)_{u_i}$$

where, for use *u*, *q* is quality value, *i* is intensity value, and *e* is exclusivity value. Therefore, each species' use value is the total of all ($q \times i \times e$) calculations for each use. This index allows a more in-depth analysis of species' usefulness than the discrete categories used by other authors. However, like all subjective allocation techniques it relies entirely on a posteriori subjective decisions by the researcher about the relative importance of species, and it is unlikely that the concept of "quality" especially would be applied consistently by different researchers.

Stoffle et al. (1990) modified Turner's model, to estimate cultural significance as a function of the number of uses, and the intensity, exclusivity, and contemporaneity of use. As they were concerned to evaluate cultural priorities in the face of development threats, they summed the use values of all plants occurring in different culturally defined "land-use areas," in order to compare the areas' relative significance to native cultures. The technique apparently has advantages and disadvantages similar to those just described for Turner's study.

Uses Totaled

No attempt is made to quantify the relative importance of each use. The number of uses (or "activities") are simply totaled, by category of plant use, plant taxon, or vegetation type. When used as an exploratory data analysis technique this approach can be extremely helpful. However, results need to be treated with caution because of two principal problems. First, the relative importance of individual uses or species is not differentiated by this technique. Second, the results are not weighted by the amount of sampling effort, so the quantity of useful plants reported can be as much an artifact of research effort as a true reflection of reality. There are, however, ways to circumvent these difficulties, as illustrated by some of the studies cited previously (see Three General Approaches to Analyzing Quantitative Ethnobotanical Data: Uses Totaled).

Both problems are arguably less important for large ecosystem-wide or countrywide compilations of uses (Bye, 1995; Moerman, 1991; Toledo et al., 1992) than for smaller-scale studies. In the former, the sheer quantity of uses reported may overwhelm any lack of resolution in an individual uses' importance, and most relevant uses may have been "captured." Few studies, however, report on how data quality and quantity vary as a function of research effort, so it is not always clear whether·a complete inventory of all culturally significant uses has been recorded.

Almost all the studies listed in Table I for the uses totaled method involved comparisons between vegetation types and categories of use or between categories of use. If the sampling effort made for each vegetation type or use category can be standardized, then these comparisons are probably valid. But if the research effort is not spread evenly, any comparisons made within a study may be questionable. To illustrate this point, imagine an ecological study in which the researcher counts the number of tree species in two different-sized sample plots: any comparison between the plots would be highly questionable if the area and number of trees sampled in each were unknown! Moreover, the degree to which findings from one study can be compared with those from others is also highly questionable (see Phillips et al., 1994: 231–232, for a case study comparing overall forest usefulness evaluated by the informant consensus and uses totaled approaches).

Conclusions

The main purpose of this chapter has been to help ethnobotanists select appropriate quantitative techniques for analyzing data on the knowledge of plant uses. However, the case for quantification is not just an academic one. Quantitative approaches may not only invigorate the science of ethnobotany, they also may help it have greater impact on wider issues of conservation and cultural survival. For example, they may help affirm the value of traditional cultures in the face of mounting external threats. In at least one example, native people and consultant quantitative ethnobotanists have recommended conservation priorities in the face of threats posed by burial of industrial waste (Stoffle et al., 1990). Also, as the debate between conservationists and economists about natural resource valuation intensifies, ethnobotanists need to be able to provide convincing information about the significance of traditional uses. Quantitative behavioral and interview ethnobotanical data can be used to help economists ascribe monetary value to subsistence uses (Brown, 1994; Chopra, 1993; Godoy et al., 1993; Pearce, 1993) or, conversely, ascribe subsistence value to commercial production (Phillips et al., 1994). In either case, the result is to help provide a common yardstick to evaluate the relative importance of different kinds of subsistence and commercial production. When economists appraise the impact of alternative development scenarios, they are less likely to ignore the subsistence value of current land use systems if standardized quantitative ethnobotanical information is available.

Acknowledgments

Miguel Alexiades, Walter Lewis, and several anonymous reviewers made helpful suggestions that greatly improved this survey. However, the views expressed here and any errors of interpretation are entirely my own. I acknowledge financial support from the NSF (Doctoral Dissertation Improvement Award BSR-9001051), WWF-US/Garden Club of America, and the Plant Program of Conservation International. I also thank the late Alwyn Gentry for support through his Pew Scholarship in Conservation and Environment during the preparation of my thesis.

Literature Cited

Adu-Tutu, Y. Afful, M., K. Asante-Appiah, D. Lieberman, J. B. Hall & M. Elvin-Lewis. 1979. Chewing stick usage in southern Ghana. Economic Botany **33:** 320–328.

Anderson, A. 1990. Extraction and forest management by rural inhabitants in the Amazon estuary. Pages 65–85 *in* A. B. Anderson, ed., Alternatives to deforestation. Columbia University Press, New York.

———. 1991. Forest management strategies by rural inhabitants in the Amazon estuary. Pages 351–360 *in* A. Gomez-Pompa, T. C. Whitmore & M. Hadley, eds., Rain forest regeneration and management. UNESCO, Paris.

——— **& D. A. Posey.** 1989. Management of a tropical scrub savanna by the Gorotire Kayapó of Brazil. Advances in Economic Botany **7:** 159–173.

Angels Bonet, M., C. Blanché & J. Valles Xirau. 1992. Ethnobotanical study in River Tenes valley (Catalonia, Iberian Peninsula). Journal of Ethnopharmacology **37:** 205–212.

Balée, W. A. 1986. Análise preliminar de inventário florestal e a etnobotânica Ka'apor (Maranhao). Boletim do Museo Paraense Emílio Goeldi **2:** 141–167.

——— **& D. Daly.** 1990. Resin classification by the Ka'apor Indians. Advances in Economic Botany **8:** 24–34.

——— **& A. Gely.** 1989. Managed forest succession in Amazonia: The Ka'apor case. Advances in Economic Botany **7:** 129–158.

Bennett, B. C. 1992. Plants and people of the Amazonian rainforests. BioScience **42:** 599–607.

Berlin, B. 1992. Ethnobiological classification: Principles of categorization of plants and animals in traditional societies. Princeton University Press, Princeton, N.J.

———, **D. E. Breedlove & P. H. Raven.** 1966. Folk taxonomies and biological classification. Science **154:** 273–275.

———, ——— & ———. 1974. Principles of Tzeltal plant classification. Academic Press, New York.

Bishop, Y. M. M., S. E. Fienberg & P. W. Holland. 1984. Discrete multivariate analysis. MIT Press, Cambridge, Mass.

Boom, B. M. 1985. Amazonian Indians and the forest environment. Nature **314:** 324.

———. 1989. Use of plant resources by the Chacobo. Advances in Economic Botany **7:** 78–96.

———. 1990. Useful plants of the Panare Indians of the Venezuelan Guayana. Advances in Economic Botany **8:** 57–76.

Brown, K. 1994. Approaches to valuing plant medicines: The economics of culture or the culture of economics? Biodiversity and Conservation **3:** 734–750.

But, P. P., S. Hu & Y. Cheung Kong. 1980. Vascular plants used in Chinese medicine. Fitoterapia **51:** 245–264.

Bye, R. 1995. Ethnobotany of the Mexican dry tropical forests. Pages 423–438 *in* S. H. Bullock, H. A. Mooney & E. Medina, eds., Seasonally dry tropical forests. Cambridge University Press, Cambridge.

Carneiro, R. L. 1978. The knowledge and use of rain forest trees by the Kuikuru Indians of central Brazil. Pages 202–216 *in* R. I. Ford, ed., The nature and status of ethnobotany. Anthropological Papers No. 67. Museum of Anthropology, University of Michigan, Ann Arbor.

Chopra, K. 1993. The value of non-timber forest products: An estimation for tropical deciduous forests in India. Economic Botany **47:** 251–257.

Elvin-Lewis, M., J. B. Hall, M. Adu-Tutu, Y. Afful, K. Asanti-Appiah & D. Lieberman. 1980. The dental health of chewing-stick users of southern Ghana: Preliminary findings. Journal of Preventative Dentistry **6:** 151–159.

Friedman, J., Z. Yaniv, A. Dafni & D. Palewitch. 1986. A preliminary classification of the healing potential of medicinal plants, based on a rational analysis of an ethnopharmacological field survey among Bedouins in the Negev Desert, Israel. Journal of Ethnopharmacology **16:** 275–287.

Godoy, R. A., R. Lubowski & A. Markandaya. 1993. A method for the economic valuation of non-timber tropical forest products. Economic Botany **47:** 220–233.

Johns, T. & E. K. Kimanani. 1991. Test of a chemical ecological model of the origins of medicinal plant use. Ethnobotany **3:** 1–10.

———, **J. O. Kokwaro & E. K. Kimanani.** 1990. Herbal remedies of the Luo of Siaya District, Kenya: Establishing quantitative criteria for consensus. Economic Botany **44:** 369–381.

———, **E. B. Mhoro, P. Sanaya & E. K. Kimanani.** 1994. Herbal remedies of the Ngorongoro District, Tanzania: A quantitative appraisal. Economic Botany **48:** 90–94.

Joly, L. G., S. Guerra, R. Séptimo, P. N. Solis, M. Correa, M. Gupta, S. Levy & F. Sandberg. 1987. Ethnobotanical inventory of medicinal plants used by the Guaymi Indians in western Panama, Part I. Journal of Ethnopharmacology **20:** 145–171.

Kainer, K. A. & M. L. Duryea. 1992. Tapping women's knowledge: Plant resource use in extractive reserves, Acre, Brazil. Economic Botany **46:** 408–425.

Kapur, S. K., A. K. Shahi, Y. K. Sarin & D. E. Moerman. 1992. The medicinal flora of Majouri-Kirchi forests (Jammu and Kashmir State), India. Journal of Ethnopharmacology **36:** 87–90.

Kroeber, A. L. 1920. Review of uses of plants by the Indians of the Missouri River region, by Melvin Randolph Gilmore. American Anthropologist **22:** 384–385.

Lee, R. B. 1979. The !Kung San: Men, women and work in a foraging society. Cambridge University Press, London.

Magurran, A. E. 1988. Ecological diversity and its measurement. Croom Helm, Kent, U.K.

McNeely, J. A., K. R. Miller, W. V. Reid, R. A. Mittermeier & T. B. Werner. 1990. Conserving the world's biological diversity. World Bank, Washington, D.C.

Moerman, D. E. 1978. Symbols and selectivity: A statistical analysis of Native American medical ethnobotany. Journal of Ethnopharmacology **1:** 111–119.

———. 1991. The medicinal flora of native North America: An analysis. Journal of Ethnopharmacology **31:** 1–42.

———. n.d. Two coins: An analysis of the food plants and drug plants of North America. Unpublished manuscript.

Paz y Miño C., G. , H. Balslev, R. Valencia R. & P. Mena V. 1991. Lianas utilizadas por los indígenas Siona-Secoya de la Amazonia del Ecuador. Reportes Técnicos **1.** Ecociencia, Quito, Ecuador.

Pearce, D. W. 1993. Economic values and the natural world. MIT Press, Cambridge, Mass.

Perez Salicrup, D. R. 1992. Evaluación de la intensidad de uso de árboles de la selva húmeda en dos comunidades de la region de Los Tuxtlas, Veracruz. Thesis. Universidad Nacional Autónoma de México, Mexico.

Phillips, O. L. & A. H. Gentry. 1933a. The useful woody plants of Tambopata, Peru. I: Statistical hypotheses tests with a new quantitative technique. Economic Botany **47:** 33–43.

——— & ———. 1993b. The useful woody plants of Tambopata, Peru. II: Further statistical tests of hypotheses in quantitative ethnobotany. Economic Botany **47:** 15–32.

———, ———, C. Reynel, P. Wilkin & C. Gálvez-Durand B. 1994. Quantitative ethnobotany and Amazonian conservation. Conservation Biology **8:** 225–248.

Pinedo-Vásquez, M. D. Zarin, P. Jipp & J. Chota-Inuma. 1990. Use-values of tree species in a communal forest reserve in northeast Peru. Conservation Biology **4:** 405–416.

Popper, K. R. 1963. Conjecture and refutations: The growth of scientific knowledge. Harper and Row, New York.

Prance, G. T. 1991. What is ethnobotany today? Journal of Ethnopharmacology **32:** 209–216.

———, **W. Balée, B. M. Boom & R. L. Carneiro.** 1987. Quantitative ethnobotany and the case for conservation in Amazonia. Conservation Biology **1:** 296–310.

Romney, A. K., S. C. Weller & W. H. Batchelder. 1986. Culture as consensus: A theory of culture and informant accuracy. American Anthropologist **88:** 313–338.

Salick, J. 1992. Amuesha forest use and management: An integration of indigenous forest use and natural forest management. Pages 305–332 in K. H. Redford & C. Padoch, eds., Conservation of neotropical forests: Working from traditional resource use. Columbia University Press, New York.

Stoffle, R. W., D. B. Halmo, M. J. Evans & J. E. Olmsted. 1990. Calculating the cultural significance of American Indian plants: Paiute and Shoshone ethnobotany at Yucca Mountain, Nevada. American Anthropologist **92:** 416–432.

Toledo, V. M., A. I. Batís, R. Bacerra, E. Martínez & C. H. Ramos. 1992. Products from the tropical rain forests of Mexico: An ethnoecological approach. Pages 99–109 in M. Plotkin & L. Famolare, eds., Non-wood products from tropical rain forests. Conservation International, Washington, D.C.

Trotter, R. T. & M. H. Logan. 1986. Informant consensus: A new approach for identifying potentially effective medicinal plants. Pages 91–112 in N. L. Etkin, ed., Plants in indigenous medicine and diet. Redgrave, Bedford Hills, New York.

Turner, N. J. 1974. Plant taxonomic systems and ethnobotany of three contemporary Indian groups of the Pacific Northwest (Haida, Bella Coola, and Lillooet). Syesis **7:** Supplement 1.

———. 1988. "The importance of a rose": Evaluating the cultural significance of plants in Thompson and Lillooet Interior Salish. American Anthropologist **90:** 272–290.

Unruh, J. & J. Alcorn. 1988. Relative dominance of the useful component in young managed fallows. Advances in Economic Botany **5:** 47–52.

10

Behavioral Orientations toward Ethnobotanical Quantification

Stanford Zent

Instituto Venezolano de Investigaciones Científicas

Introduction
Epistemology of the Etic Behavioral Research Paradigm
   Mental versus Behavioral Distinction
   Emic versus Etic Distinction
   The Emic Research Perspective
   The Etic Research Perspective
   Complementarity of Emic and Etic Approaches
   Methodological Considerations in Etic Behavioral Research
      Defining Behavioral Units
      Ensuring Representativeness
Etic Behavioral Research Methodologies
   Spatial Distribution Analysis
      Landscape Mapping
      Extrapolating Resource Production through Spatial Analysis
      Spatial Distribution of Resource Production and Productivity
   Human Activity Studies
      Time Currency
      Time and Motion Studies
      Time Allocation Studies
   Resource Accounting

Selected Guidelines for Ethnobotanical Research: A Field Manual, 199–239
Edited by Miguel N. Alexiades
© 1996 The New York Botanical Garden

Introduction

Ecological anthropology and ethnobotany are kindred scientific endeavors. The former concerns the study of the biological and cultural relationships between human communities and their natural environment. Plants make up a huge, often dominant, part of human–occupied and –managed ecosystems, hence the obvious affinity to ethnobotany. Despite the common interests, there has been relatively little interchange of theory and method between the two fields. Many of the methods currently employed in ecological anthropology are applicable and indeed useful for ethnobotanical research. Consequently, this chapter is devoted to outlining different anthropological methodologies of potential utility for the ethnobotanical researcher. The methodologies discussed here share an explicit behavioral research orientation that emphasizes the direct observation and measurement of behavioral interactions between people and plants. The behavioral focus addresses an imbalance seen in much of current ethnobotany. Whereas the field of ethnobotany has readily incorporated rigorous linguistic and cognitive-based ethnoscientific methodologies developed by anthropologists (see Berlin et al., 1974), the behavioral methodologies used by ethnobotanists are primarily descriptive and confined to laundry lists of plants and their respective uses. A behavioral approach to ethnobotany stresses systematic sampling of people's behavior with plants, quantitative data collection, and hypothesis testing through statistical analysis, leading to more microscopic and empirical as well as more macroscopic and theoretical understandings of ethnic-botanical relationships. Moreover, the quantitative study of human behavior with respect to plants is designed to capture the complexities and patterning of such behavior within the wider cultural ecological context.

It is argued here that one of the primary applications of a behavioral orientation to ethnobotany is the development of behavior-based quantitative definitions of the significance of plants in different cultural contexts. Quantitative modeling of the cultural significance or use value of plants is a recent trend in ethnobotanical research (Adu-Tutu et al., 1979; Johns et al., 1990; Phillips & Gentry, 1993a,b; Prance et al., 1987; Turner, 1988; Phillips, Chapter 9, this volume), although most of this work has focused on the quantification of *knowledge* or *ideas* about plant use or significance rather than on *actual use* patterns. The concept of activity significance, a behavioral quantitative formulation of plant significance, will be introduced in the following pages as a necessary complement to the previous quantitative treatments of cultural significance.

Epistemology of the Etic Behavioral Research Paradigm

Inasmuch as ethnobotany is concerned with the relationship between humans and plants it differs from the more conventional plant sciences precisely because of its focus on the human element. It follows that the principal objects of ethnobotanical study, human beings, are different from those in the nonhuman botanical fields because they are conscious subjects, with thoughts of their own that may be communicated via human language and behavior to the researcher. The dual nature of humans as both subjects and objects of scientific inquiry requires that certain ontological and epistemological distinctions be made explicit. The fundamental distinctions discussed here are mental versus behavioral and emic versus etic.

Mental versus Behavioral Distinction

Human biocultural activity can be divided into two broad classes of phenomena: (1) behavioral activity, which refers to physical body motions and the environmental effects of such motions, and (2) mental activity, which designates the thoughts and feelings experienced within the mind. In practical terms, behavioral and mental processes, connected through the central nervous system, are closely intertwined and to some extent interdependent, but ontologically they are separate and require that we use different kinds of research operations in order to make scientific statements about them. To describe human mental experiences, the operations must fulfill the criteria of intersubjectivity—that is, provide access to what people are thinking about. By contrast, descriptions of body motions and their environmental effects are entirely possible from an external vantage point (Harris, 1979: 30–31).

Emic versus Etic Distinction

An epistemological distinction can be drawn between the cognitive systems of the observed and those of the observer, commonly referred to as emic and etic world views, respectively (Pike, 1967). An emic point of view corresponds to the perceptions, nomenclature, classifications, knowledge, beliefs, rules, and ethics of the local plant world as defined by a native of the local cultural community. Emic knowledge allows a native person to behave in culturally appropriate and meaningful ways in different cultural contexts. An etic perspective denotes the conceptual categories and organization of the ethnobotanical environment according to the researcher, who often is an alien of the local culture and whose conceptual system ideally derives from the language and rules of science. The distinction between emic and etic kinds of plant knowledge has been most systematically treated by Berlin (1973, 1992). An important finding of his research is that there is a high level of correspondence between folk (i.e., emic) generic biological taxa and scientific (i.e., etic) biological species. However, folk generics are distinguished on the basis of perceptually salient clusters of morphological and behavioral traits, whereas scientific species are theoretically defined by the criteria of biological reproduction and evolution (Mayr, 1982).

The Emic Research Perspective

Emic and etic epistemologies have led to different and often competing paradigms for the study of human culture, each with specific implications for ethnobotanical theory and method. An emic approach to culture, which includes ethnoscientific (Sturtevant, 1964) and interpretivistic (Geertz, 1973) schools in anthropology, seeks to describe and explain cultural patterns in terms of native categories and semantic structures. Proponents of this approach subscribe to a theory of culture as a symbolic system, as a set of ideational rules for culturally appropriate behavior (Goodenough, 1964) or as a complex web of inherited public meanings embodied in symbols and communicated through social discourse and symbolic activity (Geertz, 1973). According to these formulations, the locus of culture is the collective consciousness of the cultural community, and culture (read symbol) logically precedes and determines behavior. The objective of emic research is to understand the culture in its own unique terms as a necessary first step toward subsequent generalization. Ellen (1986) has been the most forceful proponent of this approach in the field of ethnobiology, declaring that plant and animal classifications can be fully understood only in their proper social and situational contexts.

Emic methodology relies heavily on interviews with key informants from the native culture (see Alexiades, Chapter 3, this volume). Informant statements are analyzed with an eye toward discovering both conscious and unconscious structures of cultural behavior. The result, however, is often a normative or ideal description of the culture—a description of how it ought to be rather than how it really is, since natives tend to think in terms of the rules rather than the exceptions (Johnson 1978: 28; Kaplan & Manners, 1972: 22). Moreover, emic approaches have been criticized for relying too much on a few supposedly omniscient informants while ignoring intracultural variations in cultural knowledge (Pelto & Pelto, 1975; see Boster, 1986, for an informed discussion of this issue in ethnobiological research). Recent studies in ethnobotany (Adu-Tutu et al., 1979; Kainer & Duryea, 1992; Phillips & Gentry, 1993b) have overcome some of these problems by paying attention to social variables affecting plant knowledge, systematically sampling the informant pool, and making statistical comparisons of informant responses.

The Etic Research Perspective

The etic position begins with the premise that people do not always follow the rules that culture sets for them, and hence it is better to study what people actually do rather than what they say they do. The clear emphasis here is on the study of behavioral patterns, although the investigation of mental phenomena also falls within the scope of this perspective. The etic-oriented investigator describes and explains the culture on the basis of his own observations of the behavior (including verbal behavior) of the study population and according to the semantic framework provided by science. Culture is defined etically as the learned repertory of both symbols and behavior, but the principle objective here is to classify aspects of culture in terms that permit systematic comparisons with other cultures and generalizations of cultural patterns and processes according to some theoretical program (Harris, 1979). An illustration of the etic approach to ethnobotany is found in biobehavioral perspectives of ethnopharmacology, which seek to investigate the biodynamic relationships between plant use, pharmacology, and physiological import (Etkin, 1988).

Etic data collection depends on the eyewitness observation of behavioral events, often of individual behavior, by the researcher or the recording of informant recall of such events. The task of the researcher is to segment the behavior stream into significant observable units and chains of related units (see Harris, 1964, for an explicit method). Behavioral units are frequently coded according to some preconceived scheme and quantitatively measured. The perceptual, memory, and recording limitations of a single observer limit the amount of information that can be recorded at one time, although multiple observers or mechanical aids such as cameras or tape recorders are sometimes used to help make up the deficit. The objectivity of human observers, no matter how loyal to scientific principles, has also been questioned, since scientists are no less influenced and prejudiced by their own cultural and political systems than are folk people (Dumont, 1978: 45–47). This problem points out the need for explicit, detailed reporting of research methods and conditions and for improvement in the standardization of conceptual definitions and data collection techniques.

Complementarity of Emic and Etic Approaches

The differences in emic and etic research strategies have been emphasized here, but it is also important to recognize that the two strategies are not mutually exclusive; rather they occupy two ends of a methodological continuum, and the researcher should employ both emic and etic methods whenever they advance the research objectives. The classic anthropological methodologies of interviewing—recording the responses of informants to queries about cultural topics—and participant observation—observing and recording cultural activities while participating in those activities—go hand in hand and still form the crux of ethnographic (including ethnobotanical) fieldwork.

It is absurd to think that etic data collection can proceed efficiently and accurately without some access to the emic frame of reference. First, the ethnobotanist quite often enters an unfamiliar world, and, at least at first, he or she may completely miss or misinterpret behavioral patterns if unable to speak with someone who is familiar with them. For example, early in my fieldwork with the Piaroa of Venezuela, I observed a particular task that entailed prolonged kneading of the gummy mesocarp of the fruit of *Couma macrocarpa* Barb. Rodr. The kneading process, it appeared, served to expel the seeds contained therein, and these were later gathered up, parched on a stone griddle, and eaten. I initially classified this behavior as food processing (i.e., extracting edible seeds). Later, after I had become more conversant in the local vernacular, I was told that the kneaded gum is saved and used to trap birds, so I was forced to reclassify the previous recorded kneading bouts as a combination of food processing and gum trap manufacture. Second, not infrequently, the native botanist possesses a more fine-grained knowledge of the local flora than does the Western botanist, and therefore the emic knowledge system constitutes a potential source of etic botanical information. The classic example in this case is the Hanunóo of the Philippines, who were found to discriminate more plant categories and more attributes for categorization than Western systematists for the same inventory of plants (Conklin, 1954). Third, the researcher cannot always be present to witness significant behavioral events and must rely on the participants to tell what happened after the fact, or native assistants from the study community may be trained in the research methodology and employed as data collectors (Stone et al., 1990). Fourth, participants' verbal descriptions provide cues for what to look for and what to expect in the observation of complex scenes (Pelto & Pelto, 1978). The advantage of being able to communicate with the local people, whether through learning the language or using an interpreter, cannot be overstated. The important point for the etic researcher is that emic statements be translated into etic terms through effective language translation and be compared against the researcher's own observations.

In like manner, emic-oriented research can be enhanced by etic data collection. The researcher may lack the cultural and linguistic knowledge necessary for accurate comprehension of data supplied by informants (Etkin, 1988), and therefore direct observations of human–plant interactions may be needed to interpret and clarify informant statements about such interactions. Furthermore, the accuracy and completeness of verbal accounts of ethnobotanical events can be distorted by memory limitations, lack of interest, distrust, secrecy, or a host of other reasons and therefore should be verified, whenever possible, by empirical observations by the researcher. Finally, owing to the complex nature of the ethnobotanical research domain, where the focus is the relationship between different kinds of phenomena, human and natural, semiotic and physical, social and biological, the ethnobotanical researcher trained primarily in one discipline, be it anthropology or botany or geography or another, is often obliged to solicit the scientific expertise of specialists in other fields.

Methodological Considerations in Etic Behavioral Research

The etic behavioral research orientation stresses systematic and replicable collection of empirical data, problem-oriented studies that focus on the relationships between particular cultural or ecological variables, and hypothesis testing through probabilistic-statistical analysis. There are two important methodological issues to consider when doing behavioral research: (1) defining useful behavioral units and (2) ensuring representativeness of observations (Pelto & Pelto, 1978: 104).

Defining Behavioral Units

The behavioral researcher is faced with the problem of transforming the stream of observed behavior into units that are useful for description and analysis. The recording of different behavioral units should be done according to an explicitly defined code, and it is recommended that the code be compatible (and hence comparable) with the behavioral codes used in other studies (see Johnson & Johnson, 1989, for a proposal). Quantitative measurement of behavioral units is another fundamental aspect of the behavioral methodology, deemed necessary in order to achieve more precise, reliable, comparable, and statistically testable data sets (Johnson, 1978).

Ensuring Representativeness

Because ethnobotanical behavior (like knowledge) is likely to vary within a given population, the researcher must consider how to sample a representative portion of behavior. In small populations, the sample population may reach 100% of the empirical or overall population, but these situations are rare. The random sample is the most reliable sample from a statistical point of view; the haphazard sample is the least reliable. The sample type selected, however, often is a compromise between the theoretical goals, the size and diversity of the empirical population, fieldwork conditions, and time, cost, and personnel constraints of the researcher. Three common sample techniques used by human behavioral researchers are (1) the systematic sample—every nth house on a map or name in a register, (2) the stratified sample—encompassing all the relevant subgroups (e.g., male and female, rich and poor, urban and rural) of a heterogeneous population proportional to their total numbers, and (3) the cluster sample—dividing the research site into equivalent geographical compartments and subcompartments (e.g., counties, towns, blocks). One of the main causes of deviation from randomness or representativeness is the simple refusal of people to participate in the research. Another sampling problem frequently encountered is the difficulty (and danger) of attempting to observe closely members of the opposite sex of the host community. In such cases, male researchers may have to focus their data collection on men, and female researchers, on women. These are unavoidable hazards of doing research among human beings, and it is strongly advised that the researcher adapt the sampling method to local norms and desires rather than attempt to impose a method that may turn out to alienate host support and cooperation. In any case, it is important that the researcher report honestly and fully any deviations that occur in the sampling process, since these may affect the confidence in statistical outcomes.

Etic Behavioral Research Methodologies

Various behavioral methodologies developed and used by anthropologists and cultural geographers to investigate human-ecological relationships may serve the more circumscribed research interests of the ethnobotanist. Four major methodological orientations of potential application to ethnobotanical research are identified and reviewed here: spatial distribution analysis; human activity studies; resource accounting; and input-output analysis. These four methods are summarized in Table I and discussed in detail below. Such methodologies are highly relevant for developing a quantitative behavioral notion of the cultural or economic significance of plants. The behavioral approach offers a more direct form of quantifying plant utility, since human–plant interactions are measured in terms of real actions and not just words. Furthermore, a statistical understanding of people's use of and impact on plants provides a necessary control for assessing mere verbal descriptions of their relationships with plants.

Another important application of the behavioral methodologies discussed here is that they enable systematic investigation of the dynamic relationships between ethnobotanical variables and other cultural and ecological variables of the environment. Thus, it may be necessary to look at impinging social or economic factors in order to explain the origin or function of a particular ethnobotanical practice (e.g., why locally made baskets are im-

portant commodities in the local exchange system among the
Yekuana Amerindians of Venezuela even though basketmaking
knowledge and raw plant materials are evenly distributed; see
Hames & Hames, 1976). In a similar vein, plant use strategies
interact closely with other economic activities in integrated resource production systems. For example, many shifting cultivators in tropical regions manage and exploit old fallows as hunting
grounds, since game are attracted to the many plant foods found
in these areas (Nations & Nigh, 1980; Posey, 1983). A better
understanding of the systemic context of plant use may in turn
lead to a better understanding of the causality and functioning of
that use (Johns et al., 1990).

Spatial Distribution Analysis

Quantitative analysis of the spatial relationships between human
and plant communities represents an important aspect of behavioral ethnobotanical research. Three aspects of spatial distribution
analysis are discussed here: landscape mapping, estimating resource production, and spatial patterning of plant resource use.

Landscape Mapping

Landscape mapping ranges in scale, from whole culture areas or
ecotypes through the land range of a community to planting
zones within a single garden, and in level of mapping technique,
from space-based remote sensing through aerial photography to
ground survey. The scale and level chosen are directly dependent
on the research problem. In most community-based studies
where remote and macroscale data are used these must be interpreted against microscale data collected on the ground during
fieldwork (a process referred to as "ground truthing"), and in
general a "multistage" approach (i.e., integrating different levels
of data acquisition) is recommended. The increasing availability
and sophistication of Geographic Information Systems (GIS)
computer programs (e.g., ARC/INFO, IDRISI) has facilitated
this type of research by permitting fast, accurate overlay and integration of multiple spatially referenced data sets and statistical
analysis based on the correlation of spatial distributions (Aronoff,
1991; Conant, 1990).

(text continued on page 215)

Table I. Summary of four etic behavioral research methods

Method	General description and purpose	Kinds of information	Uses of information	Advantages	Disadvantages
Spatial distribution analysis	Describe and explain spatial relationships between human and plant communities	See below	Various descriptive and analytical operations dealing with spatial relationships between human and plant communities		
Landscape mapping Remote sensing	Record surface features of geographical areas via distant mechanical devices	Aerial photos, radar, Landsat or SPOT images	Macroscale analysis of synchronic or diachronic relationships between people and plants	Enables researcher to extend scope of analysis to wider area	Very costly, need special data analysis skills and equipment
Ground mapping	Map relevant natural and man-made features of local habitation and resource exploitation areas using on-the-ground techniques	Original small-scale maps of the field study site	Microscale analysis of spatial parameters of resource production, productivity, & procurement	Easy to learn, captures detail	Very time-consuming
Extrapolating resource production	Extrapolate household or community resource production from sample counts of resource production and areal measurements of production areas	Measurements and calculations of crop yields and field areas	Estimate crop production or land use intensities in fallow cultivation systems	Fast way to measure total resource production	Indirect measure of production, does-not account for crop yield or loss

Method	Description	Data	Purpose	Advantages	Limitations
Spatial distribution of resource production and productivity	Quantify differential botanical resource activities, outputs, and productivities according to geographical coordinate information	Integration of landscape mapping, human activity, and resource output data	Determine resource ranges and areas, co-variation of plant resource production and productivity with space-related variables (e.g., distance, biotope)	Facilitates various analytical operations relating to the spatial dimension of plant use	Requires accurate knowledge of the spatial location of resource activities and outputs; a good landscape map is a prerequisite
Human activity studies	Record the time spent at various plant-related behaviors through systematic observation techniques; compare the time spent at different activities	Quantitative record of how long certain activities last or how time is allocated	Statistical description and analysis of activity patterns of a community; necessary component of input-output studies		
Time and motion studies (continuous behavior sampling)	Record the duration of specific activities and subactivities by continuous observation from start to finish of the activity sequence	Database of continuous time measurements according to specific tasks	Assess the amount of time it takes to perform certain tasks; can be used to estimate time allocation	Detailed record of complete activity sequences; relatively noninvasive	Difficulty of recording groups of people; very time-consuming for researcher
Time allocation studies (state behavior sampling)	Measure the allocation of time across a range of activities by systematic scan sampling of people's behavior	Database of discrete time measurements for a range of activities that may be cross-referenced with other spatial, temporal, and social variables	Describe the pattern of time investment in work and other activities; test the co-variation of time allocation with other spatial, temporal, and social variables	Representative, accurate, and comprehensive record of a group's activities; can be economical of researcher's time	Can be quite invasive; underrecords duration of complete activities; overrecords group activities; inefficient among small and dispersed populations

Table I. Summary of four etic behavioral research methods (*continued*)

Method	General description and purpose	Kinds of information	Uses of information	Advantages	Disadvantages
Resource accounting	Keep records of resource types and amounts procured or utilized by the study community during a given period	Quantitative database of resource inventories	Derive measures of the importance of different plant species and the level of exploitation pressure on these resources; necessary component of input-output studies		
Dietary survey	Measure food intake using a variety of specific techniques	Quantitative database of food consumption habits	Assess nutritional composition and adequacy of the diet		
Weighed inventory	Weigh beginning food inventories, food imports, and ending food inventories			Economical for researcher and relatively non-intrusive for subjects	Need to control for waste amounts, exports, and consumer numbers
Dietary recall	Ask subjects to list all types and quantities of food eaten during the previous time period			Easy to administer and allows population representativeness	Highly susceptible to memory or reporting errors

Food frequency	Record the number of times a food is consumed during a given time period		Quick way to document food consumption patterns for a large group of people	Gives no quantity information about food amounts eaten within meals
Weighed intake	Weigh food at time of serving and waste amounts left over		Most accurate and precise way to measure actual food consumption	Time consuming for researcher and very intrusive for subjects
Marketing survey	Measure monetary income realized from certain activities or resource types	Account of economic return or monetary value of resources or activities	Assess income levels of people, economic performance and potential of resource types	
Ethnopharmacological survey	Observe medicinal plant uses and biomedical effects	Preclinical observations or test results of plant consumption or use and their effects	Test the efficacy of treating ailments according to local use patterns; may precede biochemical analysis or clinical testing	
Input-output analysis	Cost-benefit type of analysis of different activities using time allocation and resource accounting data	Integration of time allocation and resource accounting data; descriptive or analytical statistics	Describe or explain the interactive relationships between populations and resources	

Table I. Summary of four etic behavioral research methods (*continued*)

Method	General description and purpose	Kinds of information	Uses of information	Advantages	Disadvantages
Rational choice models	Assess economic production decisions within an environment of constraints and encouragements	Covariation of production strategies and socioeconomic variables; formal models (e.g., production function, decision trees)	Determine the influence of socioeconomic variables on production decisions; frequent tool of analysts of technological and social change	Useful in complex, differentiated social environments	primarily used in the study of agrarian situations
Optimal foraging analysis	Test optimizing assumptions of observed foraging behavior by way of mathematical maximization models	Mathematical modeling and hypothesis testing	Determine the optimal diet or resource patch mix	Theoretically and methodologically sophisticated; statistical hypothesis testing	Confined to hunting-gathering populations; requires rigorous time allocation and resource accounting data collection
Linear program analysis	Manipulate goal and constraint variables of mathematical maximization models to determine the limiting factors in an empirical situation	Mathematical modeling and hypothesis testing	Determine the degree to which empirical situations depart from optimizing assumptions	Computer analysis and simulation encouraged; mathematical-statistical hypothesis testing	Difficult to incorporate nonnumerical variables (e.g., risk, taste) into the model

REMOTE SENSING Remote sensing refers to the operation of using distant mechanical sensors (e.g., cameras, radar) to record variations in the way earth surface features reflect and emit electromagnetic energy. The primary data of remote sensing are pictorial or numerical type images taken of geographical areas of the world, such as aerial photographs or Landsat images. These images are analyzed using various viewing and interpretation devices (for pictorial data) or computers (for numerical sensor data) in order to extract information about the type, extent, location, and condition of the various resources over the area being studied (Lillesand & Kiefer, 1979: 2; see also Colwell, 1983).

The significance of remote sensing for ethnobotanical research lies in the fact that it permits the simultaneous mapping of human population and plant community distributions, which in turn sheds light on human–plant interactions. Pertinent study topics include the synchronic-spatial correlations between floristic types or even specific plant resources and human activity patterns and the diachronic impact of human activity on vegetation. Examples of remote sensing applications in ethnobotanical and related research include the interdependent relationship between cattle, grasslands, and goat herding in western Kenya (Conant, 1990); the pattern and extent of deforestation in the Amazon (Skole & Tucker, 1993); the distribution, interrelation, and evolution of land use types in a complex agricultural system in the Philippines (Conklin, 1980).

Remote images are usually acquired from conventional aircraft and from satellites such as the Landsat and SPOT series of spacecraft. For the individual researcher in the United States interested in consulting satellite images, many of these materials can be purchased from the U.S. Geological Survey at Denver, Colorado, or the private EOSAT company (formerly a division of NASA) located in Beltsville, Maryland (O. Huber, pers. comm.). The main problems with this research technique are the very high cost of producing or purchasing remote images, the technical requirements of interpreting the images, and the technical limits of the images themselves (e.g., cloud cover interferes with passive images; radar-sensed images are not collected on a regular basis; Conant, 1990).

GROUND MAPPING Ethnobotanists have traditionally carried out small-scale, intensive studies of the plant knowledge and use of a single community or group of related communities. Local maps depicting resource locations and areas and land use types are instrumental for quantitative ethnobotanical research because they facilitate various analytical operations relating to the study of plant resource production and productivity, spatial patterns of resource procurement and activity, and population–resource relationships.

Small-scale maps are usually produced using conventional ground-mapping techniques. Remote images are potentially useful here (see Conklin, 1980; Vogt, 1974), but they work best in combination with ground mapping because ethnobotanically significant microenvironmental variation does not appear on some maps and must be surveyed on the ground. Elementary surveying techniques (McCormac, 1985; Spier, 1970) can be used to produce accurate maps depicting relevant features of the field study site, including residential structures, trails or roads, prominent natural features (e.g., streams), and resource areas. The essential tools of this work include a Brunton or Suunto compass for determining directions; tape measure, pedometer, measuring wheel or rangefinder for determining distances; topographic map for reference; and grid paper and ruler for plotting data. Areal extents of resource regions, such as agricultural plots, are determined by the common perimeter method, which entails taking directional readings with a compass and measuring the distance between different points lying along the perimeter of the plot. The biggest drawback of this work is that surveying and map making are very time-consuming. However, the growing availability and affordability of global positioning system (gps) instruments and supporting computer software are helping to reduce the workload of this task.

Extrapolating Resource Production through Spatial Analysis

In agrobotanical studies, the quantification of field areas has been a key technique for extrapolating crop production amounts (Ruddle, 1974; cf. Ruthenberg, 1980). The standard methodology is to count or weigh crop portions in carefully measured areas and then multiply these figures by the total area under crop. The resulting figures give a rough estimate of crop production; further measurements must be introduced to estimate losses due to pests, intrafield variations in crop yield, or crop amounts actually harvested. Measurement of field areas has also been useful for calculating the intensity of land rotation under fallow cultivation systems (Ruthenberg, 1980). In another application of this technique, areal measures of crop production under polycultures versus monocultures were taken to determine the differential land use efficiencies of the two farming systems (Natarajan & Willey, 1980).

Spatial Distribution of Resource Production and Productivity

The determination of the spatial distribution of plant resource production and productivity is an important function of small-scale mapping in ethnobotanical research. Most people inhabit heterogeneous environments made up of different biotopes or microenvironments distinguished according to soil, vegetation, climate, and faunal characteristics. The local environment comprises a network of resource patches (i.e., clumped resource areas) in the sense that the different biotopic areas vary according to resource items, resource productivities, and distances from the settlement. At least some of these biotopes are usually man-made (e.g., gardens, secondary forests, rice paddies). Settlement pattern is to some extent conditioned by the distribution and use patterns of vegetational resource patches within a territory (see Balée, 1988; Fuentes, 1980). Seen in this light, such information might be of interest to the ethnobotanist whose research aim is to investigate the impact of the botanical environment on human sociocultural behavior.

Resource production and productivity can be calculated in terms of specific localities within the resource range, specific biotopes, or simple distance from the settlement site (Ellen, 1982: 160–168). A knowledge of simple locality is theoretically uninteresting in itself but is useful to the extent that locality is an index for other information such as area, distance, habitat, elevation, and so on. An excellent example of measuring resource production by biotope is found in Nietschmann's (1973) study of the Miskito of eastern Nicaragua. He recorded all meat biomass captured in over 40 different biotopes (many of which comprised distinct vegetational associations) over the course of a year. A more indirect approach to measuring biotopic resource productivity is to conduct censuses of the plant resources in sample plots located in different biotopes and compute the respective resource densities (Unruh & Alcorn, 1987; Unruh & Flores Paitán, 1987;

Zent, 1995). Ohtsuka (1977) studied the intensity of resource appropriation by distance among the Oriomo of New Guinea by dividing the territory of the study community into a grid of 1-km^2 blocks and recording the number of visits to each block according to different activity types.

Human Activity Studies

The observation of human activities is the most direct and simplest method for collecting data about people's interactions with plants. Anthropologists have developed and employed a number of quantitative observational techniques for studying human activity patterns that are quite useful for the ethnobotanist who seeks a quantitative understanding of people's plant-related behavior. Numerous research scenarios are conceivable: a study of the economic potential of indigenous crafts in which it may be important to know the labor costs of making different artifacts, a time series analysis of the processing techniques used to detoxify certain plant foods (Uhl & Murphy, 1981), an assessment of agricultural crop scheduling decisions (Stone et al., 1990), a study of energy expenditure of different tasks in a production system (Montgomery & Johnson, 1976).

Time Currency

Time is the most common currency or unit used by researchers for measuring human activity, although important alternative currencies include energy and money (Johnson, 1978). Time has been the favored currency because of its universal relevance, simple measurability, wide comparability across cultures, and convertibility into other currencies. The significance of time as an index of ecological behavior stems from the fact that it is a limited, and hence strategic, resource to be allocated among alternative behavioral options with different outcomes and consequences for satisfaction of biological needs. Thus, the decision to invest time in one behavior rather than another reflects an economic choice for optimizing somatic effort to produce benefit. A relevant example would be the decision to switch from a Musa- to a Manihot-dominated horticultural system by downriver Yanoama living next to mission settlements in the Upper Orinoco River of Venezuela. The shift to the more land-productive, yet

labor-expensive, manioc cultivation appears to have been motivated by the rising costs of foraging for forest foods under the
more sedentary lifestyle (Colchester, 1984). Two kinds of time
studies of human activities are discussed here: time and motion
studies and time allocation studies.

Time and Motion Studies

Time and motion studies refer to operations in which the researcher records and times an individual or group engaged in a
specific cultural activity, say, cutting a patch of forest or weaving
a basket, usually from start to finish of the activity sequence. The
time of said activity is measured in relation to a second variable,
such as resource amount harvested or area worked, in order to
be able to project how much time it takes to accomplish certain
task amounts. This technique has been used to break down work
investment in different phases of garden labor (Conklin, 1975;
Rappaport, 1968) and food processing (Uhl & Murphy, 1981;
Zent, 1992). It is equally useful for the study of other ethnobotanical events, say, preparing herbal medicines or tapping rubber.
    Time and motion studies have been adapted to experiments
testing variables of the work situation, such as differences in
technology used. Carneiro (1979) compared the work efficiency
of steel versus stone axe technology in the task of chopping
down trees by Yanomami woodcutters. The results of this study
showed that the time needed to fell a tree using a stone axe was
10 times as great as the time it took to fell a comparable tree
with a steel axe. The inevitable conclusions are that steel probably intensified horticultural patterns among tropical forest swidden farmers and, inversely, that the foraging lifestyle was a more
attractive alternative during the stone age (Colchester, 1984;
Zent, 1992). Experimental time and motion studies of this kind
are useful for examining past relationships, and, combined with
paleobiological investigations, they offer interesting possibilities
for reconstructing resource utilization patterns in prehistory.

Time Allocation Studies

More recent time allocation studies, also referred to as behavior
sampling, utilize more sophisticated methodologies for sampling
and measuring human activity patterns. The focus here is on the
distribution of time across a range of activities rather than the

duration of particular activities. Some of these methods empha-size strictly observational procedures as they were originally de-veloped for the behavioral study of nonverbal animals (see Alt-mann, 1974). Applied to the study of human conduct, methods of behavior sampling have been modified somewhat to take ad-vantage of the verbal capacities of human subjects. Many time allocation studies in fact utilize a combination of direct observa-tions of behavior and informant recording or recall of behavior.

A variety of observational methods have been proposed, and the one chosen depends on the theoretical questions, sample size requirements, length of time available for the study, complexity of the study population, and constraints of the field situation. Hames (1992: 211–212) proposed a fourfold typology of time al-location methods based on cross-cutting combinations of two di-mensional features: the duration of the observation and the num-ber of persons under observation. The dimension of duration contrasts state with continuous observations. A state observation is the recording of a single moment in time (akin to a snapshot photograph), whereas a continuous observation refers to obser-vation over a length of time (a roll of videotape). Thus in a study of basket making, the data of state observations would take the form of a frequency distribution indicating the number of times the different tasks that make up the basketmaking process (e.g., collecting raw material, drying fiber, cutting fiber, dyeing fiber, weaving) were recorded, whereas the data of continuous obser-vations would consist of a record of blocks of time spent in the different stages. The dimension of person-number distinguishes between individual and group observations. In individual sam-pling, also referred to as focal person observation, the observer records the behavior of a single individual. Group sampling, oth-erwise known as scan sampling, implies that the observer records the behavior of all persons coinhabiting an activity space. Thus, the four observational possibilities are group state behaviors, in-dividual state behaviors, group continuous behaviors, and indi-vidual continuous behaviors.

GROUP STATE BEHAVIOR SAMPLING   The observation of group state behaviors, also known as instantaneous scan sampling, spot-checks, and point sampling, is the time-sampling method most frequently used by students of human behavior (Borgerhoff Mulder &

Caro, 1985; Gross et al., 1979; cf. Gross, 1984; Hames, 1978; Johnson, 1975; Rogoff, 1978). The basic technique consists of series of observations at randomly selected time points of random or stratified samples of individuals of the study population. Time is usually randomized according to hour or half-hour slots and, in studies where time constraints limit the total days available for making observations, by day as well. The theoretical minimum period for carrying out such a study is one year in order to capture a full annual cycle of activity. The subject sample can be randomized (or stratified) by household or hamlet, and the sequence of individuals to be observed within a particular observation scan should also be randomized. The observer records the activity being performed by the subject at the moment he or she is first sighted. Borgerhoff Mulder and Caro (1985) suggest a binary code for recording and classifying behavior: (1) physical description of body movements (e.g., gathering leaves) and (2) description by consequence, amounting to a functional description (e.g., food preparation) (see also Hames, 1992). Other helpful information that should be recorded in the spot observation (besides the time, subject identification, and activity) include the date, location, weather, coparticipant(s) of the activity, and whether the subject sights the observer first (since this may cause the subject to alter his or her behavior). If someone selected for observation at a certain time is absent from view, the recorder either attempts to locate the missing person within a certain time frame or asks persons who are present where the missing person is and what he or she is likely to be doing. These "substitute" observations should be verified through follow-up interviews with the person in question when he or she shows up (Hames, 1992; Johnson, 1975). The accumulation of spot observations over a period of time (preferably a year) builds up a copious and fairly reliable quantitative data base reflecting the division of labor (and leisure) across temporal variables (e.g., hour, season) and social variables (e.g., sex, age, social class). Drawbacks of this approach include inefficiency when dealing with small and dispersed populations (Stone et al., 1990), biases related to subject's reactions to being observed and to observer reliability (Borgerhoff Mulder & Caro, 1985), bias toward overrecording group activities (Hawkes et al., 1987), unrecording of duration of complete activities and transitions and linkages between behaviors (Hames, 1992), and the lack of standard codes for classifying behaviors (but see Johnson & Johnson, 1989). Furthermore, a method that requires the researcher to search out people unannounced, indeed when they are least expecting it, can be quite invasive. Therefore the method should be adapted to the local norms of social conduct (e.g., nightly visits may not be appropriate) and used only after clearly explaining the procedure and obtaining permission from one's subjects (see also Alexiades, Chapter 1, this volume).

INDIVIDUAL STATE BEHAVIOR SAMPLING   Focal person observation is a more intensive method for recording subject behavior. The observer follows a single individual and continuously records his or her every move or, alternatively, records the subject's behaviors at given intervals, every 5 or 10 or 30 minutes. This methodology provides the most accurate and detailed accounting of time use but requires enormous investments of time by the researcher to build up a decent sample size and is extremely intrusive for the subject. The most appropriate scenario for employment of this method would be dispersed, solitary activities (e.g., plant collecting) or complex behaviors (e.g., interpersonal transactions).

CONTINUOUS BEHAVIOR SAMPLING   The method of continuous scan is analogous to the methodologies used in time and motion studies. Although this method is probably the least invasive of time allocation data collection techniques, a major problem with it is the difficulty of observing and recording all the activities of a group of performers at the same time as well as the implied high time costs for the researcher.

Resource Accounting

Resource accounting refers to the recording of resource types and amounts acquired or utilized during a specified period. This methodology is designed to produce a numerical account of the resource outputs observed among the study community. This kind of information is of obvious interest to the ethnobotanical researcher who wishes to formulate measures of the importance of different plant species to the local population and of the level of exploitation pressure on these resources. In turn, such measures facilitate comparison of plant use across cultures.

Although the method of resource accounting sounds fairly straightforward, there are actually variable procedures for counting resources. Since resources are defined in reference to particular needs, which may be defined by the research focus (e.g., food, fuel, shelter, medicine, commercial), there are different levels and currencies of resource accounting. Thus, resources may be counted at the level of resources simply acquired or at the level of resources utilized for a specific purpose. Different currencies may be used: weight, cash value, energy or nutritional value. Johnson (1978: 92) considers weight or size of the resource item harvested to be the more universal and versatile currency since it can easily be converted to other currencies.

Counting and Sampling Resource Amounts

The general technique of measuring resource outputs emphasizes counting, weighing, and sampling whenever possible. A common procedure used by researchers is to measure all resources brought back to the house or village. If the researcher must be absent, or harvests are hidden or consumed away from home, local assistants may be hired to do the counting or weighing. Another rather common technique is to interview people about their harvests of the day and estimate the amounts on the basis of their descriptions of resource names and amounts harvested and average weight or size values of the resource(s) as determined by a sample of actual weighing bouts.

Another solution for eliminating missing counts or weights is to systematically select, ideally by random sample, the days when resource harvests will be measured (Flowers, 1983). Systematic sampling of resource acquisitions is more economical than trying to achieve a total accounting, and, since the researcher can dedicate time more fully to the task on those days, may produce more accurate results. A sampling procedure is also useful for estimating total resource amounts consumed away from the house or village. Certain subjects and days can be chosen for focal following during which the resource amounts consumed on the spot are duly recorded. Another problem of representativeness in resource accounting studies is the length of study. Most authors recommend a year, but this may fall short especially in studies of wild plant resources. According to my accounts of wild fruit collection by the Piaroa of the Upper Cuao River, the fruiting schedule of some tree species (e.g., *Jessenia bataua* (Mart.) Burret, *Platonia insignis* Mart.) is more on the order of every 2 or 3 years.

Resource Data Collected

A typical resource account record should include, besides the resource type and amount, the date, name of person(s) who made the harvest, harvest location, and settlement location (if this is subject to change). Specimen collections are highly recommended in order to provide positive identification of the plant species (see Alexiades, Chapter 4, this volume). Some knowledge of local use patterns of resources also makes for more accurate and precise accounting records. For example, if a resource has more than one component part and more than one use, the researcher should attempt to separate measurement of the different portions. This can be done most efficiently by taking sample measurements. The same advice applies to measuring usable versus nonusable portions. For example, the fruit of *Couma macrocarpa* Barb. Rodr. can be broken down into four parts: skin (garbage), meat/juice (eaten raw), seed (eaten parched), and gum (used to trap birds). A comparison of the edible portions of fruit among different wild species shows some significant differences: *Couma macrocarpa,* 43.75%; *Jessenia bataua,* 43%; *Dacryodes* spp., 37.5%; and *P. insignis,* 14%. Thus in a study highlighting the dietary significance of wild plants, sample measurements of the component weights of all species encountered should be taken. The date of resource collection is useful for charting seasonal patterns of resource production, while the personal data on resource harvesters may reveal social patterns of resource production (e.g., composition of work groups or division of labor by sex or age). The data on harvest location provide useful information about spatial patterns of resource appropriation. Also under this section, it should be noted whether the resource was obtained by exchange or gift and, if by either, who and where was the source. Finally, it may prove useful to keep track of the resource destination, whether for immediate or delayed use, for local consumption or export.

Types of Resource Accounting Studies

Some studies have a more specialized focus and are interested only in particular domains of resources, in which case the accounting methods must be tailored to the domain in question. The methods most apt to interest the ethnobotanical researcher include dietary surveys, marketing or economic income surveys, and ethnopharmacological surveys.

DIETARY SURVEY   In dietary surveys, the researcher attempts to measure food intake and possibly to assess the nutritional composition and adequacy of the diet. Dietary surveys are relevant to the ethnobotanist to the extent that the dietary or nutritional significance of plants is being studied. Dietary surveys are either long-term (i.e., longitudinal surveys) or short-term (called prevalence or point prevalence surveys) (Jelliffe, 1966). Food intake can be measured in several ways, each with its pros and cons (Gibson 1990; Thompson et al., 1994). Some of the most common techniques—including weighed inventory, dietary recall, food frequency, and weighed intake methods—are briefly described below. Although these methods are specifically designed to measure food resources, they could just as easily be applied to other resource categories, such as medicines, building thatch, or charcoal.

The weighed inventory, or larder, technique entails weighing of all food at the study site at the beginning of the study, of all food brought into the site during the study, and of all food remaining at the end of the study. This technique is economical for the researcher and relatively nonintrusive for the subjects, but quantity data must be adjusted to account for waste and a close watch must be kept on consumer numbers and food exports.

The dietary recall, or history, method consists of a written or pictured questionnaire or interview in which a person is asked to report all types and quantities of foods eaten during a previous time period (usually 24 hours). This method is also fairly easy to administrate and enables population representativeness, but it is hindered by inaccurate or distorted memory or reporting.

Food frequency studies record the number of times a food is consumed during a given time period. The advantage of this method is that it represents food consumption patterns for large numbers of people relatively quickly, but it gives no quantity information about food in any particular meal.

The weighed intake method involves the weighing or measuring of foods at the time of preparation or time of serving and of waste amounts left over after meals. This is the only method in which the food amounts actually eaten are weighed and therefore yields the most accurate and precise accounting of consumption. However, the method is very time-consuming and can be extremely bothersome to subject families. Intrusiveness can be minimized by weighing the food immediately before or after it is prepared, not at mealtime as it is being eaten. Thus it may be necessary to estimate by sight the size portions that are consumed by individual family members. Depending on the research objectives and constraints, it may be useful to combine the different food-weighing methods—for example, 24-hour recall reports and weighed intake.

The nutritional survey is a dietary survey in which recorded food amounts per consumer are converted to corresponding values of dietary compounds and elements considered essential to human nutrition (e.g., calories, proteins, lipids, vitamins, minerals). A primary purpose of this type of study is to assess the nutritional adequacy of the diet. Nutritional values are assigned to food amounts using published food composition tables (McCance, 1991; Wu Leung & Flores, 1961; Wu Leung et al., 1972) or computer databases of nutritional assessment (e.g., USDA Nutrient Data Base for Standard Reference; see also Frank & Irving, 1992). With lesser known foods it may be necessary to collect samples and do your own laboratory analysis of their nutritional contents (Dufour, 1988; Murphy et al., 1991). Complete nutritional surveys require additional kinds of data: anthropometric measurements, clinical characteristics, biochemical data (e.g., hemoglobin), parasitological data, and health histories.

MARKETING SURVEY   The marketing, or economic income, survey is designed to measure the monetary income realized by individuals or household groups from certain activities or resource types over a specified time period, usually a year. Articles by Padoch et al. (1985) on market-oriented agroforestry by peasants of the Peruvian Amazon and by Romanoff (1992) on lowland Bolivian rubber tappers exemplify this type of survey. Another method of economic analysis entails recording the respective market values of various plant species and then estimating the real or potential market yield of a given land area and use on the basis of its yield of commercial plants (Peters et al., 1989).

ETHNOPHARMACOLOGICAL SURVEY   Studies of ethnopharmacologies, focused on the therapeutic and pharmacological properties of plants, include observation, interviewing about indigenous medicinal uses, and bioassays. Biomedical researchers have conducted preclinical tests of the physiological effects of local patterns of medicinal plant consumption or application (Hansson et al., 1986; Kristiansson et al., 1987). I am unaware, however, of any studies that provide quantitative descriptions of behavioral interactions with medicinal plants in a folk context, although it would seem that quantitative behavioral observational techniques should be highly relevant for biomedically oriented ethnobotanical investigations.

Input-Output Analysis

Input-output analysis encompasses a variety of research methodologies whose primary concern is to describe or explain the interactive relationships between populations and resources in human-managed ecosystems. Included under this rubric are rational choice (economic decision making) models, optimal foraging models, and linear programming analysis. These analytical techniques are all derived from mathematical optimization theory and are methodologies of data analysis properly speaking instead of data collection. A general goal here is to specify the ecological and economic (and sometimes social) factors that influence or determine resource decisions. Most studies of this genre attempt to understand human resource behavior in microeconomic terms, and, accordingly, the basic analytical technique amounts to a cost-benefit analysis of different behaviors as computed by resource output units divided by labor input units (e.g., resource weight produced per manhour; kilocalories produced per kilocalorie expended; cash received per cash invested). The collection of quantitative time allocation and resource accounting data are therefore prerequisites of input-output analysis.

Input-output analysis may be relevant for the ethnobotanist or economic botanist who wants to assess the efficiency or profitability of exploiting a given plant resource or of pursuing a given production strategy. Uhl and Murphy (1981) used input-output analysis to rate the energetic productivity of slash-and-burn farming at San Carlos, Río Negro, Venezuela. The results of this study indicated that a two-crop swidden cycle achieves the most efficient return on labor and is largely consistent with actual practice.

Rational Choice Models

Rational choice models have been applied mostly to the study of farmer production decisions (Barlett, 1980). The models attempt to assess the degree to which individuals or households follow economic maximization strategies within an environment of constraints and encouragements. A pertinent example is Barlett's (1977) study of agricultural decisions and socioeconomic differences in a rural community in Costa Rica.

Optimal Foraging Analysis

Optimal foraging analysis combines sophisticated mathematical maximization models with Darwinian theory in the analysis of foraging behavior (Winterhalder & Smith, 1981). The models enable the researcher to predict the optimal diet—the inventory of resources exploited at optimum profitability (i.e., maximum caloric yield per caloric cost)—and the optimal patch selection—the mix of resource areas foraged at optimum profitability—in a given environment. The models are operationalized by assigning a goal (e.g., labor minimization or energy maximization), a currency (usually energetic return), a set of constraints (e.g., technology available), and a set of options (e.g., the menu of resources or patches to be exploited). Hawkes et al. (1982) employed this approach to analyze the composition of plant and animal food resources exploited by Aché hunter-gatherers of Paraguay. This study showed that the Aché diet, though dominated by hunted animal foods, includes a few plant foods because the harvest rates of the latter are high enough to merit their inclusion in the optimal diet.

Linear Programming Analysis

Linear programming analysis uses many of the same mathematical maximization formulas as optimal foraging theory. The explicit purpose of linear program modeling is to manipulate goal and constraint variables in order to see what factors are most limiting in an empirical situation and to assess the degree to which empirical situations depart from optimizing assumptions. Using this approach, Johnson and Behrens (1982) determined that Machiguenga (Peru) food production decisions, which do not achieve optimal nutritional or energetic efficiency, apparently are modified by the qualitative factors of security, drudgery, diversity, and taste.

The Concept of Plant Activity Significance

The four behavioral methodologies reviewed here directly support the development of a biobehavioral approach to ethnobotany (see Etkin, 1988). These methods can be used to specify the biological and behavioral parameters of an ethnobotanical situation in order to generate and test hypothesis about human–plant interactions within the context of a broad vision of human ecology. The spatial parameters of plant use can reveal information about land use and territorial organization. The measurement of activity patterns with respect to plants is relevant to the description and evaluation of resource utilization strategies. The accounting of plant resource outputs details patterns of resource significance and exploitation pressure. Input-output analysis is a valuable tool for explaining why specific plant uses are practiced or why they are sustained or changed over time.

In addition to these general applications, the reviewed behavioral methodologies are specifically relevant for the development of a quantitative behavioral notion of the cultural or economic significance of plants. Previous formulations of this concept have been either nonbehavioral (Phillips & Gentry, 1993a,b), nonquantitative (Berlin et al., 1973), or both (Hunn, 1982). I propose the concept of **activity significance** as a behavior-based approach for quantifying the cultural significance of plants. The activity significance of a plant is defined as the set of all observed behavioral interactions between the human community and the plant. Because of the strong quantitative orientation of behavioral research, the concept of activity significance is naturally conducive to quantitative description. The methodologies described above provide the basic parameters of an activity significance—that is, spatial significance, input significance, output significance, and input-output significance.

The proposed concept of activity significance is illustrated in Figure 1, Plant Activity Significance Data Sheet. The example contains actual field data on the plant *Couma macrocarpa* for one calendar year resulting from my fieldwork with the Piaroa of the

```
                    PLANT ACTIVITY SIGNIFICANCE DATA SHEET


  I. General Data

      1. Plant Code or Number ID: _____fpw01_____

      2. Etic Name: _Couma macrocarpa Barb. Rodr._

      3. Emic Name: _upʰæ_

      4. Study Site: Upper Cuao River, Depto. Atures, Estado Amazonas, Venezuela

      5. Geographical Coordinates: (5°25' to 5°45'N 66°30' to 67°W)

      6. Ecological Setting: _evergreen basimontane-submontane-montane forest_

      7. Altitudinal Range: _250-1200 m.a.s.l._

      8. Ethnic Group: _Piaroa_

      9. Study Period: 07-01-85 to 06-30-86

      10. Project Name: Ethnoecology of Traditional Piaroa Subsistence


  II. Quantitative Data

      1. Spatial Significance
         a. Resource Location (fruit output only):
                i.   kwĕrãwĕ lagoon  (5°35'N 66°46'W - Cuao riverside)    111.24 kg
                ii.  caño æčɨ        (5°34'N 66°52'W - Cuao tributary)    218.22 kg
                iii. upper poñoto    (5°25'N 66°54'W - headwater stream)    2.6 kg
                iv.  caño baboto     (5°38'N 66°48'W - headwater stream)   33.94 kg
                v.   caño mærækænæ   (5°37'N 66°46'W - headwater stream)     52 kg
                vi.  caño ærõto      (5°36'N 66°50'W - Cuao tributary)     53.74 kg
                vii. lower kænæruoto (5°37'N 66°45'W - Cuao tributary)    138.72 kg

         b. Distance Interval from Settlement (fruit output only):
                i.   0-2 km    371.97 kg
                ii.  2-4 km    224.69 kg
                iii. 4-6 km     13.8 kg
                iv.  > 6 km        0

         c. Biotope (fruit output only):
                i.   garden              0
                ii.  old garden          0
                iii. secondary forest    0
                iv.  fluvial forest    544.68 kg
                v.   interfluvial forest 65.78 kg
                vi.  saxicolous          0
```

Figure 1. First page of Plant Activity Significance Data Sheet.

Upper Cuao River. *Couma macrocarpa* is one of the most important wild plant species exploited by the Piaroa.

The data sheet presented in the figure is one possibility of what a plant activity significance can look like. The design of the data sheet, in representation of the general outlines and details of

```
2. Input Significance

   a. Time (fruit only):

      i.  total acquisition time          103.9  hrs
             search/travel time                   --
             seed collection/prep time (cultivated)  NA
             planting time (cultivated)           NA
             weeding-pruning time (cultivated)    NA
             harvest/collection time              --
             transport time                       --

      ii. processing time
             preparing beverage:           11.5  hrs
             kneading gum:               158.94  hrs
             parching seed:               13.75  hrs

3. Output Significance

   a. Raw Harvest Amount:

      i.  fruit: 610.46 kg

                 period: 07/85 134.51 kg
                         08/85 291.26 kg
                         09/85 181.09 kg
                         05/86   1.6  kg
                         06/86   2    kg

                 use allocation: food (fruit nectar)  256.39 kg
                                             (seed)   <10.68 kg
                                 birdtrap (gum)        54.48 kg

      ii. sap:   0.7 kg

                 use allocation: paint mixer           0.7  kg

      iii. wood:  2 blocks

                 use allocation: woodwork (bench)      1 large block
                                 (blowgun mouthpiece)  1 small block

4. Input/Output Significance

   a.  raw fruit:     5.88 kg/manhour
   b.  edible fruit:  2.22 kg/manhour
   c.  edible seed:   <.09 kg/manhour
   d.  benchmaking:   1 bench/12.5 manhours
   e.  gum:            .21 kg/manhour
```

Figure I (continued). Second page of Plant Activity Significance Data Sheet.

the activity significance, is expected to vary according to the plant type, ecological and human setting, and research priorities and goals. For example, the reader will notice that much of the quantitative data given under spatial, input, and output domains refer to the harvest of *C. macrocarpa* fruit. The focus on fruit reflects my main interest in studying the Piaroa subsistence pattern. However, the same method for data organization could be applied to any other plant use or interaction type. Under the input domain, total data on acquisition time are reported. Since I did not use the focal person method of studying time allocation data, I am unable to break down time investment into the different acquisition phases of searching, collecting, and transporting. By contrast, a study focusing on foraging behavior might well want to capture that sort of data. My accounting of the quantity of *C. macrocarpa* wood harvested is also rather crude. I did not weigh blocks of wood because this aspect of resource exploitation did not form a priority of my research. In a study dealing with timber utilization, however, that type of data might be considered more noteworthy, in which case finer measurements of wood extraction would be called for. The important point is that the data sheet reflects the activity significance of the plant within the parameters and constraints of my research. Given the considerable cost of time to the researcher to collect this kind of data, it is necessary to focus on certain aspects of the activity significance of the plants (those central to the research objectives) or work in teams. Thus the individual researcher can collect only a limited subset of the total activity significance of a plant, but it is advised that it be an adequately sampled, systematically observed, correctly measured, and problem-relevant subset.

The value of compiling plant quantitative activity significance data is that it enables access to statistical details and patterns of the relationship between people and the plant that are unavailable or difficult to see in emic or qualitative descriptions. For example, the fruit production was dominated by fluvial forest trees, although several informants stated that interfluvial forest trees bear equally well (the data at least show that some interfluvial forest trees do bear fruit, but not at the same level of production as fluvial forest trees). The edible seeds of *C. macrocarpa* were mentioned as a common food, and on numerous occasions I observed the seeds being consumed. But about half the time (though I did not measure this), the seeds were thrown out and not utilized. The low overall productivity of seeds, as reported under input/output significance (<0.09 kg/manhour), may explain why this resource is deliberately wasted. Meanwhile, the cost of gum manufacture (0.21 kg/manhour), also shown under input/output significance, should be included in calculations of the profitability of trapping birds by gum traps. The distance interval data, under spatial significance, shed light not only on resource range but also on settlement pattern. We see that over 60% (371.97 kg) of *C. macrocarpa* fruit is harvested within 2 km of the settlement, whereas about 98% (596.66 kg) of the fruit is harvested within 4 km. But this rather limited range belies a multiple and mobile residence pattern. *Couma macrocarpa* is considered so important that Piaroa expressly change their settlement site, switching among mature settlements, old houses, and makeshift camphouses in order to be close to ripening tree stands and to be able to process the gum while in the vicinity. To the extent that people desire to stay within a certain distance (4 km) of *C. macrocarpa* during its fruiting season, one can surmise that this plant exerts a constraint on Upper Cuao settlement patterns. Emic information supplied by native informants often backs up the quantitative behavioral information. For example, Upper Cuao informants openly mentioned the habit of changing residence to afford easy access to *C. macrocarpa* fruit. But this anecdote merely shows the importance of investigating the correspondence between emic and etic type data; activity significance data can be used to evaluate the behavioral accuracy of the emic significance of a plant.

Finally, quantitative behavioral data lead to a more precise knowledge of the cultural or practical significance of plant taxa in the sense that they indicate the weighted significance of the respective taxa inhabiting a given significance domain. That is to say, the exact proportional significance of a taxon within the domain is knowable. For instance, my data (not shown on Figure 1) specify that *C. macrocarpa* fruit makes up 4.94% of all vegetable food weight and 49.35% of all wild vegetable food weight (Zent, 1992) in the Piaroa diet. Converted to caloric or nutritional values, such information provides a precise, absolute, and unambiguous etic sense of the dietary significance of this particular plant among the Upper Cuao Piaroa. These figures can be compared in rather straightforward fashion with the harvest weight amounts of other vegetable foods in order to gauge degrees of significance difference in the food domain. As a demonstration, we can compare the relative food weight significances of four wild plant species, in which we see the dominance of *C.*

Table II. Comparison of wild fruit raw harvest amounts

Plant species	Harvest amount (kg)
Couma macrocarpa	610.46
Scheelea sp.	161.70
Pouteria sp.	157.50
Jessenia bataua	94.18

macrocarpa (shown in Table II). The reasons for this rank ordering of food significance might be investigated by comparing the spatial distributions or resource efficiencies (i.e., input/output) of the different plant species. Comparisons of this kind can also be made for particular species in different cultural contexts, for example a survey of the food significance of *C. macrocarpa* in Venezuela or throughout the Amazon region.

Suggested Readings

The epistemologies of emic and etic research orientations are discussed by Harris (1964, 1979), Johnson (1978), Kaplan and Manners (1972), and Pelto and Pelto (1978). The basic principles of remote sensing are covered by Lillesand and Kiefer (1979) and ethnobotanical applications of remote mapping by Conant (1990) and Vogt (1974). Land surveying and mapping methods are explained by McCormac (1985) and Spier (1970). Time allocation methods are described and debated by Borgerhoff Mulder & Caro (1985), Gross (1984), Hames (1992), and Johnson (1978). Johnson (1978) briefly discusses resource accounting or outputs and input-output analysis.

Acknowledgments

I would like to thank Miguel Alexiades, Egleé López Zent, two anonymous reviewers, and the participants of Brent Berlin's Seminar in Ethnobiology at the University of Georgia for their critical observations of earlier manuscript versions. A special thanks goes to Barbara Price for suggesting that I write this paper.

Literature Cited

Adu-Tutu, M., Y. Afful, K. Asante-Appiah, D. Liberman, J. B. Hall & M. Elvin-Lewis. 1979. Chewing stick usage in southern Ghana. Economic Botany **33**: 320–328.

Altmann, J. 1974. The observational study of behavior. Behavior **48**: 1–41.

Aronoff, S. 1991. Geographic Information Systems: A management perspective. WDL Publications, Ottawa.

Balée, W. 1988. Indigenous adaptation to Amazonian palm forests. Principes **32(2)**: 47–54.

Barlett, P. 1977. The structure of decision-making in Paso. American Ethnologist **4**: 285–308.

———. 1980. Adaptive strategies in peasant agricultural production. Annual Review of Anthropology **9**: 545–573.

Berlin, B. 1973. The relation of folk systematics to biological classification and nomenclature. Annual Review of Ecology and Systematics **4**: 259–271.

———. 1992. Ethnobiological classification: Principles of categorization of plants and animals in traditional societies. Princeton University Press, Princeton, N.J.

———, **D. E. Breedlove, R. M. Laughlin & P. Raven.** 1973. Cultural significance and lexical retention in Tzeltal-Tzotzil ethnobotany. Pages 143–164 *in* M. Edmonson, ed., Meaning in Mayan languages. Mouton, The Hague.

———, **D. Breedlove & P. Raven.** 1974. Principles of Tzeltal plant classification: An introduction to the botanical ethnography of a Mayan-speaking community in highland Chiapas. Academic Press, New York.

Borgerhoff Mulder, M. & T. M. Caro. 1985. The use of quantitative observational techniques in anthropology. Current Anthropology **26**: 323–336.

Boster, J. 1986. "Requiem for the Omniscient Informant": There's life in the old girl yet. Pages 177–198 *in* J. W. D. Dougherty, ed., Directions in cognitive anthropology. University of Illinois Press, Urbana.

Carneiro, R. L. 1979. Forest clearance among the Yanomamö, observations and implications. Antropológica **52**: 39–76.

Colchester, M. 1984. Rethinking Stone Age economics: Some speculations concerning the pre-Columbian Yanoama economy. Human Ecology **12**: 291–314.

Colwell, R. (ed.). 1983. Manual of remote sensing. Vol. 1: Theory, instruments and techniques. Vol. 2: Interpretations and applications. American Society of Photogrammetry, Falls Church, Virg.

Conant, F. P. 1990. 1990 and beyond: Satellite remote sensing and ecological anthropology. Pages 357–388 *in* E. F. Moran, ed., The ecosystem approach in anthropology: From concept to practice. University of Michigan Press, Ann Arbor.

Conklin, H. C. 1954. The relation of Hanunóo culture to the plant world. Dissertation. Yale University, New Haven, Conn.

———. 1975. [1957]. Hanunóo agriculture, a report on an integral system of shifting cultivation in the Philippines. Forestry Development Paper 12. Food and Agricultural Organization of the United Nations, Rome.

————. 1980. Ethnographic atlas of Ifugao: A study of environment, culture, and society in northern Luzon. Yale University Press, New Haven, Conn.

Dufour, D. 1988. The composition of some foods used in northwest Amazonia. Interciencia **13(2)**: 83–86.

Dumont, J. P. 1978. The headman and I: Ambiguity and ambivalence in the fieldwork experience. Waveland Press, Prospect Heights, Illinois.

Ellen, R. 1982. Environment, subsistence and system: The ecology of small-scale social formations. Cambridge University Press, Cambridge.

————. 1986. Ethnobiology, cognition, and the structure of prehension: Some general theoretical notes. Journal of Ethnobiology. **6**: 83–98.

Etkin, N. L. 1988. Ethnopharmacology: Biobehavioral approaches in the anthropological study of indigenous medicines. Annual Review of Anthropology **17**: 23–42.

Flowers, N. M. 1983. Seasonal factors in subsistence, nutrition, and child growth in a central Brazilian Indian community. Pages 357–390 *in* R. B. Hames & W. T. Vickers, eds., Adaptive responses of native Amazonians. Academic Press, New York.

Frank, R. C. & H. B. Irving. 1992. Directory of food and nutrition information for professionals and consumers. 2nd ed. Oryx Press, Phoenix, Ariz.

Fuentes, E. 1980. Los Yanomami y las plantas silvestres. Antropológica. **54**: 3–138.

Geertz, C. 1973. The interpretation of cultures. Basic Books, New York.

Gibson, R. S. 1990. Principles of nutritional assessment. Oxford University Press, Oxford.

Goodenough, W. 1964. Cultural anthropology and linguistics. Pages 36–39 *in* D. Hymes, ed., Language in culture and society. Harper and Row, New York.

Gross, D. R. 1984. Time allocation: A tool for the study of cultural behavior. Annual Review of Anthropology **13**: 519–558.

————**, G. Eiten, N. M. Flowers, F. M. Leoi, M. L. Ritter & D. W. Werner.** 1979. Ecology and acculturation among native peoples of central Brazil. Science **206**: 1043–1050.

Hames, R. B. 1978. A behavioral account of the division of labor among the Ye'kwana Indians of southern Venezuela. Dissertation. University of California, Santa Barbara.

————. 1992. Time allocation. Pages 203–235 *in* E. Smith & B. Winterhalder, eds., Evolutionary ecology and human behavior. Aldine de Gruyter, New York.

————**& H. L. Hames.** 1976. Ye'kwana basketry: Its cultural context. Antropológica **44**: 3–58.

Hansson, A., G. Veliz, C. Naquira, M. Amren, M. Arroyo & G. Arevalo. 1986. Preclinical and clinical studies with latex from *Ficus glabrata* HBK: A traditional intestinal antihelminthic in the Amazon area. Journal of Ethnopharmacology **17**: 105–138.

Harris, M. 1964. The nature of cultural things. Random House, New York.

————. 1979. Cultural materialism: The struggle for a science of culture. Random House, New York.

Hawkes, K., K. Hill & J. O'Connell. 1982. Why hunters gather: Optimal foraging and the Aché of eastern Paraguay. American Ethnologist **9:** 622–626.

————, **H. Kaplan, K. Hill & A. M. Hurtado.** 1987. A problem of bias in scan sampling. Journal of Anthropological Research **43:** 239–247.

Hunn, E. 1982. The utilitarian factor in folk biological classification. American Anthropologist **84:** 830–847.

Jelliffe, D. B. 1966. The assessment of the nutritional status of the community. World Health Organization, Geneva.

Johns, T., J. O. Kokwaro & E. K. Kimanani. 1990. Herbal remedies of the Luo of Siaya District, Kenya: Establishing quantitative criteria for consensus. Economic Botany **44:** 369–381.

Johnson, A. W. 1975. Time allocation in a Machiguenga community. Ethnology **14(3):** 301–310.

————. 1978. Quantification in cultural anthropology: An introduction to research design. Stanford University Press, Stanford, Calif.

———— **& C. A. Behrens.** 1982. Nutritional criteria in Machiguenga food production decisions: A linear-programming analysis. Human Ecology **10:** 167–189.

———— **& O. Johnson.** 1989. Machiguenga time allocation data base: Cross-cultural studies in time allocation. HRAF Press, New Haven, Conn.

Kainer, K. A. & M. L. Duryea. 1992. Tapping women's knowledge: Plant resource use in extractive reserves, Acre, Brazil. Economic Botany **46:** 408–425.

Kaplan, D. & R. Manners. 1972. Culture theory. Prentice Hall, Englewood Cliffs, N.J.

Kristiansson, B., N. Abdul Ghani, M. Eriksson, M. Garle & A. Qirbi. 1987. Use of khat in lactating women: A pilot study on breast-milk secretion. Journal of Ethnopharmacology **21:** 85–90.

Lillesand, T. M. & R. W. Kiefer. 1979. Remote sensing and image interpretation. John Wiley, New York.

Mayr, E. 1982. The growth of biological thought: Diversity, evolution, and inheritance. Harvard University Press, Cambridge, Mass.

McCance, R. A. 1991. McCance and Widdowson's *The composition of foods*. Royal Society of Chemistry, Ministry of Agriculture Fisheries and Food, Cambridge, U.K.

McCormac, J. C. 1985. Surveying. 2nd ed. Prentice Hall, Englewood Cliffs, N.J.

Montgomery, E. & A. Johnson. 1976. Machiguenga energy expenditure. Ecology of Food and Nutrition **6:** 97–105.

Murphy, S. P., S. W. Weinberg-Andersson, C. Neumann, K. Mulligan & D. H. Calloway. 1991. Development of research nutrient bases: An example using foods consumed in rural Kenya. Journal of Food·Composition and Analysis **4(1):** 2–17.

Natarajan, M. & R. W. Willey. 1980. Sorghum-pigeonpea intercropping and the effects of plant population density. 2: Resource use. Journal of Agricultural Science **95:** 59–65.

Nations, J. & R. Nigh. 1980. The evolutionary potential of Lacandon Maya sustained yield tropical forest agriculture. Journal of Anthropological Research **36:** 1–30.

Nietschmann, B. 1973. *Between land and water.* Seminar Press, New York.

Ohtsuka, R. 1977. Time-space use of the Papuans depending on sago and game. Pages 231–260 *in* H. Watanabe, ed., Human activity system: Its spatiotemporal structure. University of Tokyo Press, Tokyo.

Padoch, C., J. Chota Inuma, W. de Jong & J. Unruh. 1985. Amazonian agroforestry: A market-oriented system in Peru. Agroforestry Systems **3:** 47–58.

Pelto, P. J. & G. H. Pelto. 1975. Intra-cultural diversity: Some theoretical issues. American Ethnologist **2:** 1–18.

————— & —————. 1978. Anthropological research: The structure of inquiry. 2nd ed. Cambridge University Press, Cambridge.

Peters, C. M., A. H. Gentry & R. Mendelsohn. 1989. Valuation of an Amazonian rainforest. Nature **339:** 655–656.

Phillips, O. & A. H. Gentry. 1993a. The useful woody plants of Tambopata, Peru: I. Statistical hypotheses tests with a new quantitative technique. Economic Botany **47:** 15–32.

————— & —————. 1993b. The useful woody plants of Tambopata, Peru: II. Additional hypothesis testing in quantitative ethnobotany. Economic Botany **47:** 33–43.

Pike, K. L. 1967. Language in relation to a unified theory of the structure of human behavior. 2nd ed. Mouton, The Hague.

Posey, D. 1983. Indigenous ecological knowledge and development of the Amazon. Pages 225–257 *in* E. Moran, ed., The dilemma of Amazonian development. Westview Press, Boulder, Colo.

Prance, G. T., W. Balée, B. M. Boom & R. L. Carneiro. 1987. Quantitative ethnobotany and the case for conservation in Amazonia. Conservation Biology **1:** 296–310.

Rappaport, R. A. 1968. Pigs for the ancestor: Ritual ecology of a New Guinea people. Yale University Press, New Haven, Conn.

Rogoff, B. 1978. Spot observation: An introduction and examination. Quarterly Newsletter of the Institute for Comparative Human Development, Rockefeller University **2(2):** 21–26.

Romanoff, S. 1992. Food and debt among rubber tappers in the Bolivian Amazon. Human Organization **51(2):** 122–135.

Ruddle, K. 1974. The Yukpa cultivation system: A study of shifting cultivation in Colombia and Venezuela. Ibero-Americana 52. University of California Press, Berkeley.

Ruthenberg, H. 1980. Farming systems in the tropics. 3rd ed. Clarendon Press, Oxford.

Skole, D. & C. Tucker. 1993. Tropical deforestation and habitat fragmentation in the Amazon: Satellite data from 1978 to 1988. Science **260:** 1905–1910.

Spier, R. F. G. 1970. Surveying and mapping: A manual of simplified techniques. Holt, Rhinehart, and Winston, New York.

Stone, G. D., R. McC. Netting & M. P. Stone. 1990. Seasonality, labor scheduling, and agricultural intensification in the Nigerian savanna. American Anthropologist **92:** 7–23.

Sturtevant, W. 1964. Studies in ethnoscience. American Anthropologist **66:** 99–131.

Thompson, F. E., T. Byers & L. Kohlmeier. 1994. Dietary assessment resource manual. Journal of Nutrition **124(11)** Supplement I–V: 2245S–2317S.

Turner, N.J. 1988. "The importance of a rose": Evaluating the cultural significance of plants in Thompson and Lilloet Interior Salish. American Anthropologist. **90:** 272–290.

Uhl, C. & P. Murphy. 1981. A comparison of productivities and energy values between slash and burn agriculture and secondary succession in the Upper Río Negro region of the Amazon Basin. Agro-Ecosystems **7:** 63–83.

Unruh, J. & J. Alcorn. 1987. Relative dominance of the useful component in young managed fallows at Brillo Nuevo. *In* W. Denevan & C. Padoch, eds., Swidden-fallow agroforestry. Advances in Economic Botany **5:** 47–52.

——— **& S. Flores Paitán.** 1987. Relative abundance of the useful component in old managed fallows at Brillo Nuevo. *In* W. Denevan & C. Padoch, eds., Swidden-fallow agroforestry. Advances in Economic Botany **5:** 67–73.

Vogt, E. Z. (ed.). 1974. Aerial photography in anthropological field research. Harvard University Press, Cambridge, Mass.

Winterhalder, B. & E. A. Smith (eds.). 1981. Hunter-gatherer foraging strategies: Ethnographic and archaeological analyses. The University of Chicago Press, Chicago.

Wu Leung, W. T., R. Batrum & F. (Huang) Chang. 1972. *Food composition tables for use in East Asia.* U.S. Department of Health, Education and Welfare. National Institute of Arthritis, Metabolism and Digestive Diseases, Bethesda, Maryland.

——— **& M. Flores.** 1961. Food composition table for use in Latin America. Institute of Nutrition of Central America and Panama and the Interdepartmental Committee on Nutrition for National Defense, Bethesda, Maryland.

Zent, S. 1992. Historical and ethnographic ecology of the Upper Cuao River Wõthĩhã: Clues for an interpretation of native Guianese social organization. Dissertation. Columbia University, New York.

———. 1995. Clasificación, explotación y composición de bosques secundarios en el Alto Río Cuao, Estado Amazonas, Venezuela. Pages 79–113 *in* H. D. Heinen, J. San José, & H. Caballero, eds., Naturaleza y ecología humana en el Neotrópico. Scientia Guaianae **5.** Caracas, Venezuela.

11

Beyond Nomenclature and Use: A Review of Ecological Methods for Ethnobotanists

Charles M. Peters
Institute of Economic Botany,
The New York Botanical Garden

Selected Guidelines for Ethnobotanical Research: A Field Manual, 241–276
Edited by Miguel N. Alexiades
© 1996 The New York Botanical Garden

Introduction

Although broadly defined as the study of the interrelationships between plants and people, ethnobotany has, in most cases, focused solely on compiling lists of the plant species used by different cultural groups. Recent studies have expanded on this concept somewhat by trying to quantify the relative importance of different plant uses (see Phillips, Chapter 9, this volume) or by focusing in greater detail on the actual pattern or intensity of use of different resources (see Zent, Chapter 10, this volume). These modifications notwithstanding, ethnobotany has remained primarily a static, descriptive endeavor. The core components of the discipline today—plant collection, plant identification, and the detailed documentation of plant uses at one point in time—are essentially the same as they were 100 years ago when Harshberger first coined the term *ethnobotany* (Harshberger, 1896).

The basic shortcoming of a purely descriptive approach is that it does not take into account the fact that things happen when people use plants. Destructive harvesting and overexploitation, for example, can gradually eliminate a plant species from the local environment. Deliberate planting, controlled harvesting, and forest management, on the other hand, can greatly increase the distribution and abundance of local resources. Species lists alone are insufficient to document these dynamic interactions. A particular plant resource may be recorded as having exceptional properties and a high use value during one ethnobotanical survey, but if the species that produces it occurs at low densities in the forest, is harvested destructively, or cannot regenerate under existing levels of exploitation, there is a very high probability that the resource will not even be noted in subsequent surveys. There is an ecological context within which people interact with plants, and the exploration of this territory can generate a host of new questions for inquisitive ethnobotanists. Perhaps it is time to go beyond the basic queries of What is the name of this plant? and

What is it used for? and to ask How do indigenous communities apply their knowledge of the local flora? and, perhaps most important, What are the long-term impacts of these actions?

The purposes of this review are to outline some of the basic ecological methods available for addressing the latter types of questions and to briefly introduce the reader to the literature on quantitative ecology and vegetation sampling. Particular attention is focused on the collection of quantitative density and yield data for different plant resources. The relative advantages and limitations of a variety of sampling procedures are discussed, and potential methodological problems are highlighted whenever appropriate. Most of the examples presented are taken from my own research in the tropical forests of Amazonia and Southeast Asia. This review is not exhaustive, nor is it a "cookbook" of ecological methods that can be applied without modification. The selection of an appropriate sampling scheme for ethnobotanical work will ultimately depend on the specific objectives of the research, the experience and judgment of the investigator, and the time, financial resources, and personnel available.

Quantitative Assessment of Species Density

Density, or the number of individuals per unit area, is probably the ecological parameter of greatest interest to the ethnobotanist. This basic statistic can tell the investigator how much of a given plant resource is available for exploitation and where the greatest abundance of harvestable material is located. If the individuals are measured as well as counted, size-specific density estimates can be obtained to assess whether the species is regenerating under exploitation. Quantitative estimates of species density also lay the foundation for ecological monitoring by providing a yardstick with which to measure the long-term sustainability of plant resource exploitation (Hall & Bawa, 1993; Peters, 1994).

For cultivated plants growing in house gardens or small agroforestry plots, it is sometimes possible to conduct a 100% inventory of all individuals to obtain a precise estimate of species density (e.g., Padoch & de Jong 1991; Rico-Gray et al., 1990). In most situations, however, it is neither feasible nor warranted to count all of the individuals of a species, and some type of sampling methodology will be required. The major issues to be considered in selecting an appropriate sampling scheme for collecting density data are related to the size, shape, number, and arrangement of sample plots. Also important, of course, are the procedures and measurements used in the field, deciding which plants to count, and what variables to measure.

Size and Shape of Sample Plots

Beyond the general advice that larger plants require larger sample plots, there are few rules that govern the selection of an appropriate plot size for vegetation sampling. Plots of 1.0 m² are usually sufficient for use with herbaceous plants (Kershaw & Looney, 1985; Van Dyne et al., 1963), but shrubs and understory vegetation may require plots of from 16.0 to 100 m² (Lyon 1968; Myers & Chapman, 1953). A variety of plot sizes have been used to sample forest vegetation. For inventory work in tropical forests, Lang et al. (1971) and Knight (1975) recommended the use of 10 × 20 m plots; 1000-m² plots were used in the comparative studies of Holdridge et al. (1971) and Gentry (1982); and various investigators have used large, single plots of from 1.0 to 3.0 hectares (e.g., Anderson et al., 1985; Campbell et al., 1986; Gentry, 1990; Hubbell, 1979). Although 100-m² plots are frequently recommended for measuring the density of tree species in temperate hardwood forests (Mueller-Dombois & Ellenberg, 1974), some studies (e.g., Bormann, 1953; Whittaker, 1967) found that 1000-m² or 1400-m² plots gave better results in these plant communities.

Logistic factors are also important to consider in selecting a plot size. Large sample plots, which have a greater probability of encountering different patches or "clumps" of individuals than do small plots, will usually provide a better estimate of the mean density of a species (Gauch, 1982). Large plots, however, take longer to lay out and inventory, and, given that there are more individuals to measure and count, there are more chances to make mistakes. Small plots are immeasurably easier to lay out and count, but they frequently produce a density estimate with a large error term, especially if the plot size selected is smaller than the average size of the natural aggregations formed by the species being surveyed (Greig-Smith, 1983). If most of the sample plots fall directly within a clump, the final density estimate will be too high. If most of the plots miss these clumps, or only partially bisect them, the density of the species will be underestimated. As can be appreciated, the selection of an appropriate plot size for density sampling represents a compromise between time, expense, and level of precision required.

In terms of overall sampling efficiency, the actual shape of the plot may be more important than its size. As is shown in Figure 1, plots of similar size can exhibit notable differences in the amount of perimeter, or "edge," depending on their shape. For a given sample area, circular plots have less edge than square plots, and square plots have less edge than rectangular plots or transects. The total perimeter of a rectangular plot is controlled by the ratio of plot width to plot length, and long, narrow transects have considerably more perimeter than short, wide ones. To illustrate the magnitude of this range, I used different plot configurations to calculate the maximum and minimum perimeter lengths for each transect area shown in Figure 1.

There are positive and negative aspects associated with the perimeter characteristics of a sample plot. On the positive side, a larger amount of edge means that the sample unit will usually bisect a greater number of different habitats and species patches and provide a more representative description of the study area. This benefit is enhanced by orienting the long axis of the transect at right angles to topographic or drainage features (Avery & Burkhart, 1983). The extensive use of transects in floristic surveys (see reviews in Campbell et al., 1986; Gentry, 1982) and forest inventory operations (FAO, 1973; Wood, 1989) is largely the result of the increased perimeter afforded by sample units with this shape.

On the negative side, plots that have a large perimeter or boundary also have a large number of boundary or border trees whose inclusion or exclusion from the sample must always be assessed. The treatment of border trees is a chronic source of error in plot sampling. Ideally, the investigator should carefully measure out from the centerline of the plot to every border tree to verify that it is actually "in." This, however, can be a very time-consuming process, and, in most cases, the distance is sim-

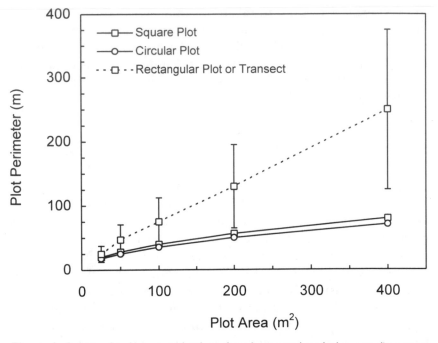

Figure I. Relationship between plot boundary (in meters) and plot area (in square meters) for sample units of different sizes and configurations. Mean perimeter length is plotted for rectangular plots; minimum and maximum lengths for each rectangular plot area are indicated by vertical bars.

ply estimated visually by the investigator. The net result of this procedural shortcut is that too many stems are usually counted near the perimeter of a plot. Other sources of subjective bias can also enter in the treatment of border trees. For example, if the person responsible for determining whether a border tree is in or out is also the one who must climb the tree to collect the herbarium specimen, there is a tendency late in the day for large canopy trees to be out and for pole-sized, easily collectible stems to be in. Similarly, there is always the subtle temptation in floristic surveys to include the border trees that represent new species and to exclude those of species that have already been tallied.

Small sampling errors due to edge effects and the careless assessment of border trees can quickly add up. Consider for the moment a 10×1000 m transect run through a lowland tropical forest. If we assume that the questionable border area of the transect is 50 cm wide, there is approximately 1000 m² of perimeter, or about 10% of the total plot area, within which a decision must be made as to which trees are in or out. Further assuming that the average density of trees in the forest is 700 stems/ha and that these stems are evenly distributed throughout the site, the investigator conducting the inventory should encounter about 70 border trees, or, stated another way, be confronted with at least 70 chances for committing a sampling error. To fully reap the statistical benefits of using rectangular plots and transects, special care should be taken to ensure that all plot boundaries are precisely maintained.

Number of Sample Plots

The total number of plots to use for collecting density data will necessarily depend on the spatial heterogeneity or patchiness of the species being sampled. Plant species that exhibit a regular or homogeneous spatial pattern can be sampled with fewer plots than species that grow in pronounced clumps. For species that are distributed at random throughout the study site, the precision of the density estimate will depend solely on the total number of individuals counted in the inventory, regardless of the size or shape of the individual sample plots used (Greig-Smith, 1983). As a general rule, the investigator should adapt a "more is better" philosophy and always try to sample as many plots as possible within a given habitat.

Several basic methods are available for determining the appropriate number of plots to use for a particular species or vegetation type. One method involves calculating the running mean, or variance, of the density estimates obtained from successive plot samples and then plotting these against the total number of plots sampled (Goldsmith & Harrison, 1976; Kershaw & Looney, 1985). In most cases, the variation in mean density will be quite high among the first plots sampled and then will gradually flatten out as the calculated sample mean begins to more closely approximate the true density of stems in the study area. An example of this technique is shown in Figure 2 using stem data collected from a series of 10×20 m contiguous plots in lowland dipterocarp forest within the Danau Sentarum Wildlife Reserve, West Kalimantan, Indonesia (Peters, unpubl. data). The curve exhibits a considerable amount of fluctuation up to a sample size of about

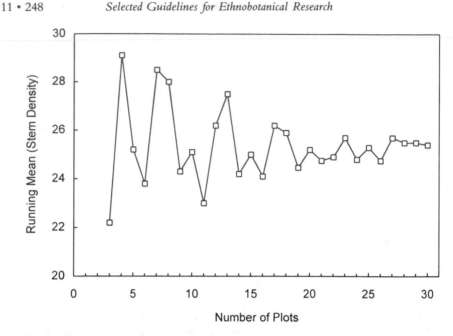

Figure 2. Running mean of number of stems (≤10 cm in diameter) per plot along a series of 30 contiguous 10 × 20 m plots. Data collected from a lowland dipterocarp forest in the Danau Sentarum Wildlife Refuge, West Kalimantan, Indonesia. (From Peters, unpubl. data).

22 plots (4400 m²), at which point the variation in mean density starts to stabilize. This pattern suggests that a minimum sample size of approximately 25 plots (0.5 ha) would be sufficient for estimating the density of stems in this forest.

In many commercial timber surveys, it is common practice to set the sample intensity as a certain percentage of the total sample area to ensure that sufficient plots are sampled. Sample percentages of from 5% to 10% are usually standard (Avery & Burkhart, 1983; Bonham, 1989). Assuming that a 5% sample percentage is desired, a 100-ha tract of community forest would need to be sampled with 50,000 m² of plots. This area could be obtained by using five 10 × 1000 m transects, fifty 20 × 50 m rectangular plots, or 100 circular plots with a radius of 12.62 m. The obvious problem with this method is that not all species and vegetation types require the same sample percentage. Blindly using a constant percentage in all situations will cause the investigator to use too many plots in some cases and not enough plots in others.

Several statistical techniques have been developed for estimating the exact number of sample units needed to obtain a given level of precision. These calculations, however, require an a priori estimate of the mean and variance of the population to be measured from a pilot survey or from a few preliminary plots that have been sampled in the study area. Detailed discussions of these techniques can be found in Husch et al., 1972; Köhl, 1993; and Philip, 1994. Most of these techniques require that the data be collected using a random sampling design (see below).

Arrangement of Sample Plots

A final consideration of great importance in the quantitative assessment of species density has to do with the way in which the plots are arranged throughout the study area. There are essentially two methods for deciding where to locate the sample units: the samples may be distributed regularly throughout the area in a systematic fashion, or they may be randomly located. A third method, the subjective placement of sample plots in typical or representative sites, is usually not recommended. Subjective sampling carries with it an excessive degree of personal bias, and the data collected using this procedure are not acceptable for any statistical tests involving the assessment of significance, such as t-tests, regression, correlation, or F-tests (Cochran, 1977; Greig-Smith 1983).

In systematic sampling, the sample units are spaced at fixed intervals. The location of the first sample is usually selected at random and all other samples are positioned according to a strict pattern. The actual sample units employed may be either transects or plots. Figure 3 illustrates the general layout of a systematic transect sample (Figure 3A) and a systematic plot sample (Figure 3B.). The square sample area shown is equivalent to a 100-ha tract of forest composed of three different forest types or land use categories (I, II, and III).

The transects shown in Figure 3A are 10 m wide and 1000 m long. Note that the transects are parallel and that they have been oriented at right angles to the rivers (solid black lines) so that all soil types and environmental conditions are intersected to provide a representative sample of the local vegetation. The 200-m interval between transects results in an overall sample intensity

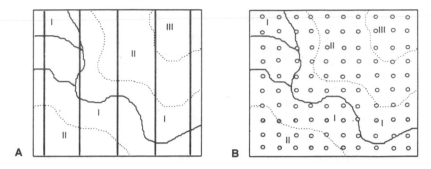

Figure 3. Arrangement of sample units in systematic sampling design. Square area shown represents 100-ha tract composed of three different forest types or land use categories (I, II, III). Solid black lines represent rivers; dotted lines represent type boundaries. **(A)** Systematic transect sample. Transects are 10 m wide, 1000 m long, and separated by 200 m, resulting in an overall sample intensity of 5%. **(B)** Systematic plot sample. Plots are circular, with a radius of 12.62 m, yielding a sample area of 500 m². Total sample intensity equals 5% (i.e., 100 plots × 500 m² plot = 50,000 m², or 5.0 ha).

of 5%. The percent sampling intensity of a systematic transect sample is calculated simply by dividing the transect width (10 m) by the distance between transects (200 m) and then multiplying by 100. This feature is extremely useful in cases where the boundaries of a forest tract are known but the total area has yet to be determined. In the present example, a total of 5000 m of transect, or 50,000 m², were sampled. Given a sample percentage of 5%, the total area of the tract, if unknown, could have been calculated by multiplying the reciprocal of the sampling percentage (20) by the total sample area (50,000 m²) to give a result of 1,000,000 m², or 100 ha.

The basic design of a systematic plot sample is shown in Figure 3B. The plots are uniformly spaced throughout the forest in a grid pattern along north–south (10 columns) and east–west (10 rows) compass bearings. The circular configuration, with a 12.62-m radius (500 m²), of each of the 100 sample plots reduces the number of boundary trees and edge. The total sample area obtained by this design is 50,000 m² (100 plots × 500 m² per plot), resulting in the same sample intensity (5%) as that provided by the transects. Given the even coverage provided by the grid layout, the plots can be oriented without worrying about the topography or drainage features of the site.

Forest inventories based on systematic sampling present several distinct advantages relative to other sample designs. First, they provide a good estimate of population means and totals because the sample area is spread out over the entire study site. Second, they are faster and less expensive to conduct than randomized designs because the location of the sample units is based on fixed directional bearings and distances. Locating transects or plots in the field is greatly facilitated, and travel time between sample units is minimized. Third, because the entire site is traversed in a regular, controlled pattern, supplementary forest type or land use information can be collected and easily mapped during field operations. Finally, systematic sampling does not require a priori knowledge of the total area of vegetation to be sampled.

A systematic design, however, has one major disadvantage. There is no satisfactory way to estimate the precision or sampling error of the data collected, because statistical variance computations require a minimum of two randomly selected sample units (Grieg-Smith 1983; Husch et al., 1972). In systematic sampling, only the location of the first plot or transect is selected at random; the remaining sample units follow a predetermined and regular pattern. This would not be a problem if all of the trees in a housegarden, agroforestry field, or managed forest were distributed at random and exhibited no pattern of variation. Unfortunately, the individuals in a biological population are rarely, if ever, arranged independently of each other, and there is a high degree of natural variability. It is, therefore, impossible using systematically collected data to separate the variability attributed to randomness from that naturally exhibited by the population. Although worthy of note, this limitation detracts very little from the overall utility of systematic sampling. In practice, the lack of an estimable sample error means only that the density data from two different areas cannot be compared statistically.

A random sampling design, on the other hand, provides not only mean and total density values but also an estimate of the precision of those values. The calculation of a standard error ($s_{\bar{x}}$) and confidence limits (CL) from the sample data, for example, allows the investigator to state that, at any given probability level, the true density value for the population or species lies within a certain specified range (Snedecor & Cochran, 1967).

Two examples of random inventory designs are illustrated in Figure 4. As before, the total area shown is 100 ha. There are 100 circular sample plots (each with a radius of 12.62 m), and the sample intensity is 5%. In simple random sampling (Figure 4A), all 100 plots are randomly located throughout the study area. A convenient method for determining the location of random plots is to place a transparent grid over a base map, aerial photo, or satellite image of the area and to draw randomly generated pairs of Cartesian coordinates for each plot. After marking the location of all plots on the base map or photo, derive compass bearings and distances from a central starting point to describe their relative position in the field.

The example shown in Figure 4B represents a stratified random sampling design. Although at first glance the pattern seems identical to that of simple random sampling, the important difference is that the plots have been "stratified" by forest type. The number of sampling units allocated to each forest type is determined by the percentage of the total area represented by each type, so that larger forest types contain a greater number of plots. Besides providing a more precise and efficient sample design, stratification also helps to avoid the uneven distribution or

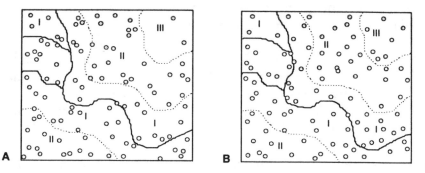

Figure 4. Arrangement of sample units in random sampling design. Square area shown represents 100-ha tract composed of three different forest types or land use categories (I, II, III). Solid black lines represent rivers; dotted lines represent type boundaries. All (100) plots are circular, with a radius of 12.62 m (total sample area 500 m²); plot location is based on randomly selected coordinates. Overall sample intensity is 5%. **(A)** Simple random sample. Plots are located randomly throughout the entire area. Note that forest type III (upper right corner of figure) is sampled by only two plots. **(B)** Stratified random sample. The number of plots allocated to each forest type is based on percentage of total area represented by that type. Note that forest type III has now been sampled with seven plots

clumping of plots that frequently occurs with random sampling (Adlard, 1990; Philip, 1994). In Figure 4A, for instance, a large area of type I forest remains unsampled. The same procedure used to locate plots for simple random sampling can be used for stratified random sampling. Random coordinates that place a plot within a forest type requiring no further sampling are simply rejected.

The clear advantage of simple random sampling is its statistical rigor. Precise confidence limits can be assigned to all of the data, and, given information on the natural variability of the population (e.g., from a preliminary sample), the minimal number of plots that need to be used to adequately describe the forest can be calculated. There are, however, several disadvantages to a random design. The plots can be very difficult to locate in the field, and much time is wasted traveling from one plot to the next. In some cases, the random selection of plot location may leave significant sections of the study area unsampled. Perhaps the greatest limitation, however, is that random sampling does not allow the regular, grid-based observations necessary for detailed forest type or land use mapping.

Both random and systematic sampling are routinely used to collect plant density data. Wood (1989) reported that systematic sampling is the preferred design for commercial timber inventories in Africa and Southeast Asia, whereas random sampling is more strongly favored in Latin America. Of the 36 tropical countries surveyed in Wood's study, fixed-area plots had a higher frequency of use (44%) than transects (34%). In terms of ethnobotanical research, systematic (Kinnard, 1992; Lepofsky, 1992) or subjective (Balée, 1994; Boom, 1989; Prance et al., 1987) transects seem to be the preferred sampling method, although some investigators have opted for a random (Pinedo-Vasquez et al., 1990) or stratified random (Irvine, 1989; Salick, 1989) design. Single 1.0-ha sample plots have also been used in several recent studies (e.g., Fong, 1992; Phillips, 1993).

Field Procedures and Measurements

Regardless of the sampling design selected, great care should be taken in locating and laying out the sample plots in the field. If the plots fall in the wrong place, or are the wrong size, it is very likely that the final density estimate obtained will also be wrong. Of special importance in this regard is the correction for slope. Ensuring consistency and comparability requires that all distance and area measurements be made along the horizontal. Measuring 20 m along a 10% slope yields a very different horizontal distance (19.9 m) than measuring 20 m along a 40% slope (18.6 m). In both cases, however, the horizontal distance obtained is less than that desired. In terms of geometry, measuring along a slope is like measuring the hypotenuse of a right triangle, when what the investigator should really be trying to measure is the base of the triangle.

Failure to correct for slope can lead to significant measurement errors. Take, for example, the case of a 10 × 1000 m transect laid out along a constant 30% slope. If no correction is made for the topography, every 20 m measured along the transect will be 0.8 m too short and each 10 × 20 m segment of the transect will contain 192 m^2, rather than 200 m^2. By the end of transect, the sample unit will be 40 m too short and will contain 400 m^2 less than it should. The density data from this sample are clearly not comparable to those collected from a 10 × 1000 m transect laid out along flat terrain.

Table I shows one way to avoid this problem. Slope corrections are tabulated for different distances and percent slopes. As indicated in the table, the measurement of 10 horizontal meters along a 30% slope requires a distance of 10.44 m. Meter tapes or ropes can be prepared in advance to facilitate the use of these correction factors in the field. If measurements are made at 10-m intervals, for example, a 15-m tape or rope can be marked or knotted at the appropriate distances for the range of slopes expected to be encountered in the field. One crew member pulls the tape 10 m, while a second stays behind to take a slope reading (e.g., with a clinometer or Abney level). If the slope was determined to be 50%, then the tape would be extended to the 11.2 m mark before setting a plot stake or tallying the next 10-m segment of the transect.

Which Plants to Count?

Two points in particular should be addressed in deciding which plants to count in each of the plots. The first is related to the

Table I. Slope corrections for different distances and percent slopes. Table values indicate the distance along a slope that must be traveled to obtain the horizontal distance indicated by the column heading.

Slope (%)	Horizontal distance (m)				
	5	10	15	20	25
10	5.02	10.05	15.07	20.10	25.12
15	5.06	10.11	15.17	20.22	25.28
20	5.10	10.20	15.30	20.40	25.50
25	5.15	10.31	15.46	20.62	25.77
30	5.22	10.44	15.66	20.88	26.10
35	5.30	10.59	15.89	21.19	26.49
40	5.39	10.77	16.16	21.54	26.93
45	5.48	10.97	16.45	21.93	27.41
50	5.59	11.18	16.77	22.36	27.95
55	5.71	11.41	17.12	22.83	28.35
60	5.83	11.66	17.49	23.32	29.15
65	5.96	11.93	17.89	23.85	29.82
70	6.10	12.21	18.31	24.41	30.52
75	6.25	12.50	18.75	25.00	31.25
80	6.40	12.81	19.21	25.61	32.02
85	6.56	13.12	19.69	26.25	32.81
90	6.73	13.45	20.18	26.91	33.63
95	6.90	13.79	20.69	27.59	34.48
100	7.07	14.14	21.21	28.28	35.36

minimum size limit of the individuals to be included in the sample. The second has to do with deriving an operational definition of the word *individual*. Both of these issues should be resolved before fieldwork is begun.

The minimum size limit used in an inventory exerts a controlling influence on the total number of plant stems that have to be counted. In most cases, the smaller the minimum size limit, the greater the number of sample plants that are included. To illustrate this relationship specifically for trees, inventory data from 1.0-ha samples of hill dipterocarp forest and managed forest orchard in West Kalimantan, Indonesia, are shown in Figure 5. The stem counts from these samples (Peters, unpubl. data) were grouped into 5.0 cm diameter classes. In both environments, the increase in the number of sample trees is linear down to a diameter of about 25–30 cm. Further decreases in minimum diameter,

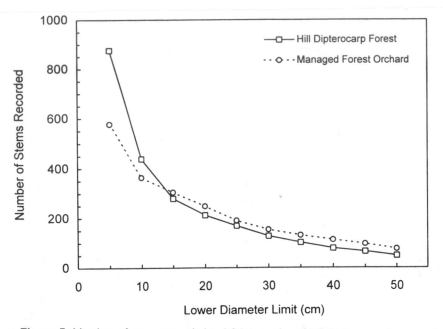

Figure 5. Number of stems recorded in 1.0-ha samples of hill dipterocarp forest and managed forest orchard using different minimum diameter limits. Data for hill dipterocarp forest were collected in the Raya-Pasi Nature Reserve; managed forest orchard data are from the Dayak village of Bagak Sahwa. Both sites are located in the Sambas district of West Kalimantan, Indonesia. (From Peters, unpubl. data.)

however, result in an exponential increase in the number of stems; the density of 5.0-cm sample trees ($n = 876$) is almost twice that of 10.0-cm trees ($n = 439$) in the unmanaged forest. The lower number of small stems in the managed forests reflects the periodic thinning and selective weeding practiced by local Dayak communities (Padoch & Peters, 1993).

The pattern shown in Figure 5 underscores the trade-offs involved in choosing a size limit for inventory work. A smaller minimum size increases the amount of information obtained from the sample, but it also greatly increases the time and expense of fieldwork. Larger diameter cut-offs significantly speed up field operations, but they may result in an unrepresentative sample of certain plant resources. Many important forest fruits, for instance, are understory and midcanopy species, and these resources would be completely missed by adopting the 20–40 cm diameter limit used in many commercial timber surveys (Heinsdijk & de Bastos, 1965; UNESCO, 1978). As a compromise between time invested and information obtained, a large number of ethnobotanical studies (e.g., Phillips, 1993; Pinedo-Vasquez et al., 1990; Prance et al., 1987) and floristic surveys (e.g., Campbell et al., 1986; Gentry, 1988) have used a 10-cm minimum diameter limit. Studies focused specifically on the collection of density data for acaulescent palms, lianas, bamboos, or herbaceous plants would, of course, require the use of an even smaller minimum size limit.

For most tree species, it is relatively easy to distinguish one genetic individual, or *genet,* from another. They originate from a single seed, produce a single trunk, and occupy a well-defined and exclusive space in the forest. The situation is a bit more complicated for bamboos, caespitose palms, and many herbaceous plants that reproduce vegetatively and have the ability to form dense, multistemmed clumps. In these species, a single genetic individual may be represented by innumerable clonal shoots, or *ramets.* Faced with an impenetrable stand of bamboo, or a rattan clump with over 200 spiny stems, or a dense sward of grass with innumerable tillers, the investigator may be hard pressed to even distinguish where one individual stops and another begins. What should be counted—the individual clumps (genets), the individual stems (ramets), or both?

Probably the best rule of thumb in these cases is to try to quantify the same vegetative unit as that which is actually exploited as a resource. If an entire sward of grass (i.e., one genet composed of many ramets), for instance, is pulled up and used as thatch, counting the number of clumps per plot will probably yield a reasonable estimate of the density of this resource. Bamboo poles (i.e., ramets), which are harvested individually, would best be assessed by counting all of the culms in each plot. In some cases, however, it is clearly not feasible to count all of the individual ramets produced by a species (e.g., a dense cluster of rattan or an extensive bamboo forest). Counting all the clumps (genets) in the plot and then choosing a subsample of individuals of varying size for counting ramets is a useful strategy for getting around this problem. Once all of the plot data have been collected, regression analyses can be used to derive a predictive equation describing the relationship between genet size and number of ramets. Based on the total number of clumps counted in the plots, this equation can be used to provide a rough estimate of the total number of harvestable rattan canes or bamboo stems growing in the study area.

What to Measure?

Although the collection of density data is essentially a counting process, there is much to be gained by measuring the size of each of the sample individuals encountered in the plots. For example, it is useful to know that each hectare of your study area contains 100 fruit trees >10 cm dbh. Determining that 40 of these trees are over 30 cm in diameter and of sufficient size to actually produce fruit is a finding of even greater utility. Information about the size distribution of individuals in a population can also frequently provide indirect evidence about the regeneration success of that species. A population composed of 100 trees per hectare with all of the individuals over 50 cm in diameter is very different from one composed of 100 trees per hectare with 20 trees 50 cm in diameter, 30 trees of 30 cm dbh, and 50 trees of 10 cm dbh. The latter population appears to be regenerating itself quite well. The former is probably destined to disappear from the site as soon as the big trees die. The really important question here is not so much What is the total density of species A within the study area? as What is the density of different-sized individuals [i.e., the size structure] of species A on the site? Answering this question requires that the sample plants be measured as well as counted.

The most frequently measured and easily obtainable expression of tree size is diameter at breast height, or dbh (approximately 1.4 m above the ground). In cases where extreme buttress formation, wounds, or forked boles preclude the measurement of dbh, a section of clear trunk immediately above the problem area should be measured, and a note that this was done should be recorded near the diameter measurement for that individual (Philip, 1994). Depending on the resource, dbh may not always be the most meaningful size parameter to measure. Basal diameter (i.e., at the ground level) is a more useful index to classify the size structure of shrub populations (e.g., Peters & Vazquez, 1987; Reid et al., 1990), and height, although difficult to measure with precision in closed forest, may be the only alternative for members of those taxa (e.g., palms, bamboos, and tree ferns) that do not grow in diameter (Ash, 1987; Bullock, 1980; Pinard, 1993; Piñero et al., 1977).

Growth and Yield Studies

An ethnobotanist assessing the use of resources by a local population is interested in *how many* individuals of a particular species are growing in local forests, agroforestry fields, or home gardens. These density estimates, however, are only half of the plant-use equation. To really evaluate the quantity of resource available to local populations, one must also quantify *how much* of the desired resource is produced by each of these individuals. Foresters routinely collect this type of data by monitoring the radial increment of timber trees, and there is a large and detailed literature on the growth and yield characteristics of commercial timber species (e.g., Adlard, 1990; Alder, 1980; Wan Razali et al., 1989). The situation, however, is quite different for nontimber resources. Virtually nothing is known about the fruit, oil seed, latex, or resin yield of forest species or local cultivars, even for the most valuable and widely exploited market species. How many Brazil nuts does a large *Bertholletia* tree produce? How much rattan cane does a wild *Calamus* clump make in a year? What is the rubber yield from a large *Hevea* tree in the lowland forests of Amazonia? Although fertile ground for ethnobotanical inquiry, these questions remain essentially unanswered.

Selection of Sample Trees

The basic objective of a yield study is to provide a reasonable estimate of the quantity of resource produced by a given species growing in a particular habitat. As it is rarely feasible to monitor all of the individuals of a selected species, data collection will necessarily focus on a subsample of plants. If at all possible, the selection of these sample plants should be stratified by two main variables—diameter, or some other indicator of plant size, and site condition.

There are several good reasons for conducting a stratified sample. Regardless of the species or type of resource produced, plant size exerts a major influence on yield. Large plants, because of their better canopy position, larger leaf area and root mass, and greater availability of stored carbohydrates, are usually significantly more productive than smaller plants (Kozlowski et al., 1991). The actual parameter of interest, therefore, is not simply mean production but the size-specific production rate of the species. As will later be discussed in this section, the exact nature of the statistical relationship between size and yield is an important tool for deriving an estimate of the total annual productivity of a given plant resource.

Plant productivity also varies with respect to certain site parameters. Even after the effect of size has been accounted for, most species will usually exhibit higher yields in some sites than in others. The key variable of interest may be soil depth and fertility, soil moisture, canopy cover, relative slope, forest type, or presence and absence of cultural treatments such as thinning, weeding, or mulching. Whatever variables are selected, the important thing is that the investigator be able to stratify or partition the local habitat along these lines. The selection of slope as a growth parameter, for example, would require some knowledge of the areal extent of different slope conditions within the study area, as well as an estimate of the density of plants occurring in each slope class. These data, or those for soil condition, canopy cover, forest type, or management status, could be collected during the inventory work by taking the appropriate measurements in each sample plot.

There are no hard and fast rules to determine the number of sample plants that should be selected to assess the yield characteristics of a particular species. In many cases, the issue will resolve itself on the basis of the relative density of the species in different site conditions, the size distribution of the population, the number of individuals in each diameter or height class, and the actual time and expense available for conducting yield studies. As was discussed earlier in regard to the number of sample plots, the greater the number of individuals that can sampled, the greater the accuracy of the final estimate.

Using the results from the plot survey as a guide, the investigator should randomly select individuals from different size classes and habitats. If at all possible, the number of sample trees selected from each size class should be the same in each site condition or forest type. Ideally, every size class should be sampled by *at least* three individuals so that some index of variability (such as standard error) can be calculated. Size classes can sometimes be lumped together to achieve this objective if fewer than three individuals per class are available. A reasonable level of precision can be ensured if the total number of sample plants selected in each site condition falls within the range of 25–30 trees (e.g., 5–6 trees in each of five size classes).

After the sample plants to be measured have been selected, these individuals should be located in the field, sequentially numbered, and permanently labeled with paper or plastic tags. If sufficient time and funding are available, additional information such as crown area or number of leaves (for palms), canopy cover, and distance to and size of nearest-neighbor trees can also be collected from each sample individual at this time. These data can later be grouped into classes and compared statistically to provide a more detailed analysis of size-specific productivity.

Methodology and Data Collection

The exact sampling procedure used in the yield studies will necessarily vary with the type of resource being measured. For ease of discussion, the innumerable useful products produced by plants can be divided into three main groups based on the origin of the plant tissue or compound being used: reproductive propagules (e.g., fruits, seeds, and accessory tissues), plant exudates (e.g., latexes, resins, and gums), and vegetative tissues (e.g., stems, leaves, roots, barks, and apical buds). Although fruits, nuts, and oilseeds are different commodities, their production by individual trees can be measured using a similar methodology.

Reproductive Propagules

The production of fruits and seeds is measured at discrete intervals throughout the fruiting season using either direct counts or a random sample of the area under the crown of adult trees. For small trees that produce few fruits of relatively large size (e.g., shrubs, some palms, and cauliflorous trees), direct counts of fruit can be employed with reasonable precision (Dinerstein, 1986; Peters & Vazquez, 1987; Piñero & Sarukhan, 1982; Sork, 1987). It is usually a good idea to make replicate fruit counts on the same individuals until a consistent number is obtained. If nuts or seeds are the actual resource of interest and the fruits in question are multiseeded, a large number of mature fruits should be opened ($n = 50$–100) and the seeds counted to determine the average number of seeds per fruit.

Direct counts can also be used with species that produce multiple fruits in large infructescences (e.g., many palms) by harvesting these structures when mature and carefully counting the number of fruits (e.g., Anderson et al., 1985; Phillips, 1993). An alternative strategy is to first record the total number of infructescences produced by the tree and then to harvest a subsample for counting individual fruits. Given the ease of direct counts, it is tempting to use this procedure on large-fruited canopy trees by scanning the crown with binoculars. This technique, however, is not recommended as there is no way to mark the fruit that have already been counted, and it is extremely difficult to survey the entire crown of a large tree without some degree of overlap or repetition. Chapman et al. (1992), for example, compared visual estimates of fruit production by two canopy species in Uganda with actual counts of the total fruit crop collected under the crowns of the trees and found correlations of only marginal significance between the two data sets.

Tall forest trees that produce more fruit than can be counted individually must be sampled using small plots or specially constructed fruit traps (see review in Green & Johnson, 1994). A critical assumption involved in using this method is that a large percentage of the fruits will fall directly under the crown of the adult tree. For most commercial fruits, nuts, and oilseeds, which are relatively large and heavy, this assumption probably is valid. However, sampling under the crown of a tree will not account for the fruits and seeds that are eaten or dispersed by animals before they fall, and, as a result, the data collected will not represent the total number of fruits produced. This limitation notwithstanding, the use of fruit traps or plots does provide a reasonable estimate of postpredation or harvestable yield, which may actually be a more relevant and useful measurement for the ethnobotanist than is total yield.

The first step in the sampling process is to determine the exact area of the vertical projection of the crown of each sample tree. This is accomplished by measuring out from the trunk of the tree to the outermost branches of the crown along at least four radii. On the basis of these measurements, the actual projection or "shadow" of the crown is sketched on millimetric graph paper and its area calculated using the appropriate formula for that configuration (e.g., circle or ellipse). A stratified random design is then used to allocate the sample plots or traps within this area. The crown area is divided into four quadrants of similar area; the boundaries of these quadrants are determined by four perpendicular radii extending out from the truck. Random coordinates are then chosen to position the samples within each quadrant. The reason for this stratification is that fruits rarely fall in a symmetrical or regular pattern under a tree. Prevailing winds and the relative position of fruit-laden branches usually cause more fruits to fall on one side of the crown than the other. Dividing the crown projection into quadrants will ensure that regions of both high and low fruit density are sampled.

There are two options for determining the number of sample plots or traps to be used under each tree. A constant percentage of the crown area can be sampled (Howe, 1980; Howe & Vande Kerckhove, 1981), or, alternatively, a constant number of traps can be used irrespective of crown area (Howe, 1977; Peters, 1990; Peters & Hammond, 1990). The former method requires that a greater number of samples be located under large trees than small ones, and certain statistical tests may be complicated because of the unequal sample sizes (Sokal & Rohlf, 1981). The latter method samples smaller trees more intensively than large ones. If a fixed sampling percentage is desired, a sufficient number of traps should be used to sample about 10% of the total crown area. If a constant number of traps of sample units is used, a total of 8–12 traps or plots (i.e., 3–4 per quadrant) should be located under each tree.

Although traps may have a slight advantage over plots in that fruits cannot roll out of the sample unit, plots are faster to lay out and easier to maintain. The most common plot size used in fruit production studies is a 1×1 m square. The plots under each tree should be numbered sequentially, the corners staked, and the boundaries clearly delineated with plastic string or flagging. Raking the plot down to mineral soil and maintaining it free from leaves and vegetation can greatly facilitate the locating of fallen fruits.

Fruit traps can be from 0.5 to 1.0 m^2 in size and of either a square or circular configuration. The smaller traps are somewhat more stable and easier to transport to the field. Square traps are constructed by first making a box frame out of 1×4 cm wooden battens and then stretching a piece of 2-mm nylon netting tightly over the bottom and affixing it with tacks or staples (see description in Adlard, 1990). Circular traps can be made out of stiff wire or plastic tubing; larger fruits require a stronger and more durable trap than smaller ones. A 79.8 cm diameter circle has an area of 0.5 m; a 112.8 cm diameter circle provides a 1.0-m^2 sample area. Nylon screening is used to make a loose, concave net (approximately 30 cm deep), which is then tied or clipped to the circular frame; plastic clothespins work very well for clipping the bag to the hoop. Both square and circular fruit traps should be elevated about 50 cm off the ground using treated wooden stakes or PVC pipe, and the number of the trap should be clearly marked on the leg or frame.

Fallen fruits start to decompose quite rapidly on the forest floor, and there is always the possibility that some fruits will be eaten or removed by animals before they are counted. Fruit predation between sampling periods can also be a problem with traps, which are easily climbed by squirrels and other forest rodents. Reviewing the traps or plots as frequently as possible, preferably twice a week, will help prevent these potential sources of error. At each sampling period, the number of immature and mature fruits in each sample unit should be carefully counted and all of the reproductive material removed from the plot. Screens, stakes, and plot boundaries should also be checked at this time and repaired if necessary. The biweekly sampling of each tree should be continued until at least two consecutive fruit counts give null results.

Plant Exudates

The measurement of plant exudate yield requires some a priori knowledge of the traditional tapping or collection technique used for a particular species. Of special importance is information concerning the frequency with which the trees are usually tapped. Through a continual process of trial and error over the years, experienced collectors have undoubtedly determined the tapping method and harvest schedule that produces the greatest amount of latex, resin, or gum. The objective here is to actually quantify this yield.

Perhaps the easiest way to obtain these data is through careful participant observation; work with an experienced local assistant and follow him around as he taps the sample trees (see Barrera de Jorgenson, 1993; Dove, 1993; Gianno, 1986). The exudate obtained from each tree is measured (by weight or liquid volume depending on the resource) and recorded in the field, and, with the help of the local assistant, an initial estimate is made of the frequency with which the tree can be tapped. Several sequential tappings should be measured to obtain some idea of the variability in yield, as well as to observe the tree's response to the wounding caused by harvesting. Depending on the particular tapping regime employed, daily, weekly, or monthly production rates are then calculated for each sample tree and exudate under study.

Vegetative Tissues

The variety of vegetative structures exploited as resources—e.g., stems, leaves, barks, roots, apical buds—can be divided into two groups based on the physiological response of the plant species to harvesting: the plant species will either survive and later regenerate the vegetative structures removed, or it will be killed by harvesting the tissue. The former group includes leaves, branches, and the bark and apical buds of certain species; the latter includes most types of stem tissue, roots, and bark. Different sampling methodologies are required to estimate the productivity of these two groups.

For species that exhibit regrowth or sprouting, the basic idea is to first quantify the existing stock of harvestable resource and then to monitor the rate at which these resources are replenished by the plant. The periodic collection of palm leaves provides a useful example to illustrate this concept. Working with experienced collectors, the investigator records the average number of leaves harvested from the crown of each sample individual together with data on the total number of leaves per crown. The residual leaves on each individual should be marked with paint or tags to differentiate them from the new leaves that are later produced. After an adequate period of time has passed for the new leaves to fully elongate, the palm is reharvested and the leaf number is again determined. This procedure should be followed through at least two cycles of harvest and new leaf production to get some idea of whether the rate of leaf production decreases in response to repeated harvest. The mean yield figure for each tree represents the total number of new leaves harvested throughout the sample period. The final result should be adjusted to reflect a yearly production rate. Similar methodologies have been used in studies of plant demography to quantify the leaf replacement rates of palms (Bullock, 1980; Lugo & Rivera, 1987; Oyama 1990; Sarukhan, 1978), cycads (Clark & Clark, 1987), and tree ferns (Tanner, 1983).

The procedure for collecting yield data for species that are killed by harvesting is a bit more complicated. Measuring root growth or the production of bark, for example, is plagued with methodological difficulties, and these resources require a two–step sampling scheme involving allometric analyses and growth studies. During the first phase of this process, the preselected sample plants are felled, dissected, and carefully measured to obtain an estimate of the size-specific bark or root volume for that species. Regression analyses are then used to derive a predictive equation relating plant size to quantity of resource present. The slope of this regression line can eventually be used to predict yield.

The problem, however, is that there is, as yet, no time dimension or rate associated with the production of the resource. What is lacking is information about the rate at which these plants grow from one size class to the next. Collecting this information requires selection of a second subsample of plants representing a range of different sizes and habitats; growth of the subsample plants is monitored for at least one year. Diameter growth is the best parameter to measure for most tree species, and these data can be collected by using dendrometer bands (Bormann & Kozlowski, 1962; Liming, 1957) or by making periodic diameter measurements on the same sample trees. In the latter method, painting a line on each of the sample trees to indicate the original point of measurement is highly recommended. Height growth is clearly a more meaningful parameter to measure for palms, herbs, and understory plants.

Combining the data sets from the allometric and growth studies provides a reasonable estimate of the productivity of a particular root or bark resource. For example, if the bark biomass of a 20–cm *Cinchona* tree is 11.0 kg and the bark biomass of a 25–cm tree is 20 kg (Hodge, 1948), a 20–cm tree growing at 0.5 cm per year would produce approximately 900 g of bark tissue a year. If necessary, the order of the allometric and growth studies can be reversed, with the growth studies being conducted first and the same sample trees later being harvested and analyzed. This strategy, which requires only one group of sample trees, may be warranted for species occurring in low-density populations.

Solitary rattans are an especially difficult subject for yield studies. In many respects, the harvest of these resources is identical to logging in that entire stems are removed and, for many species, there is no resprouting (Dransfield & Manokaran, 1994). The problem is that there is no easily measurable indicator of growth, such as diameter (dbh), as is used by foresters to estimate productivity. Rattans, like all palms, have no secondary meristem and exhibit no radial growth. They produce new stem tissue (cane) solely by extension growth. For small and intermediate-sized individuals, height growth can be measured directly to obtain an estimate of cane yield. Measuring the height increment of the larger, more valuable canes, some of which may be 40–50 m long, is quite a bit more difficult and requires tree climbing.

There is usually no way to get around this problem. Basing yield figures solely on the extension growth of smaller, and frequently slower-growing, individuals will lead to an underestimate of productivity. Periodic controlled harvests can be used to estimate the local stock of rattan cane, but this procedure ignores the critical issue of size-specific yield. Perhaps the only recommendation that can be made is to try to measure at least a few large-sized canes. To achieve this objective, one must climb each individual to locate its apical bud or growing point. A point on the stem immediately behind the bud should then be permanently marked with paint and tied with flagging to facilitate relocation. After 6–12 months, the climber should enter the canopy again and carefully measure the distance from the paint mark to the end of the apical bud. The average growth rate taken from several large canes could then be applied to all large-sized, canopy individuals.

Defining the Resource Base

The results from the fieldwork described thus far can be integrated to estimate the total quantity of harvestable resource produced by different plant populations within the study area. These data, which represent the size of the resource base for a particular species, can be used to analyze local patterns of resource exploitation, to forecast future yields and harvest revenues, and to assess the ecological sustainability of current harvest levels. They are, in essence, the foundation upon which resource management is based.

Two pieces of information are needed for this analysis: size-specific production data from the growth and yield studies and the size structure of the population obtained from the inventory data. Although the following discussion uses rattan as an example, the basic procedure for estimating total population yield is essentially the same for all types of plant resources.

The size-specific production data collected in the yield studies are first grouped by habitat or site class and then regression analyses are performed to derive a predictive equation describing the relationship between plant size and productivity. In some cases, the functional relationship between these two variables will not be linear, and the data may require some type of transformation (e.g., conversion to logarithms) before they are analyzed (Sokal & Rohlf, 1981). It is also possible that after inspection the data could be best described using curvilinear or polynomial regression techniques. Whatever degree of analysis is used, however, the objective is to produce a result that is both biologically meaningful and statistically significant.

One example of the form that this relationship might take is shown in Figure 6 using data collected for *Calamus schistoacanthus* Bl. in the Danau Sentarum Wildlife Reserve in West Kalimantan, Indonesia (Peters, unpubl. data). This rattan is an important source of cordage and weaving material for local fishing communities, and large quantities are harvested every year for both subsistence use and sale. The species occurs naturally in high-density stands in the seasonally flooded forests of the reserve. Growth data were collected in 1994 from a total of 78 *C. schistoacanthus* individuals representing a range of height classes. As is indicated by the regression line in Figure 6, there is a linear relationship between plant size and extension growth for canes up to 5 m in length, and the relationship appears to be strong enough from a statistical standpoint ($r^2 = 0.774$; $P < 0.01$) that stem growth can be reliably predicted from cane length.

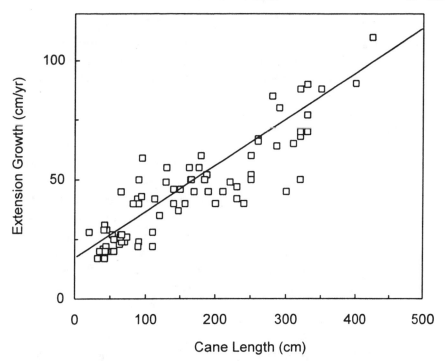

Figure 6. Annual extension growth as related to cane height for *Calamus schistoacan-thus* rattans ($n = 78$) growing in the seasonally flooded forests of the Danau Sentarum Wildlife Reserve, West Kalimantan, Indonesia. The regression line is based on the general linear model, growth $= a + b$ (height); the parameter values and coefficient of determination are: $a = 15.76$, $b = 0.186$, $r^2 = 0.78$, $P < 0.01$.

The final equations obtained from the yield studies are used to estimate the collective productivity of each of the appropriate size classes (i.e., those containing individuals of reproductive or merchantable size) in the population. These estimates are calculated by substituting the midpoint of each size class as the dependent, or y, variable in the yield equation. The average yield value for each size class is then multiplied by the actual number of individuals within that class to obtain a class total. Summing these totals over all size classes provides an estimate of total population yield. Care should be taken to include only productive individuals in these calculations. The male trees of dioecious species, for example, obviously should be omitted from an analysis of total fruit yield.

The inventory and growth data for *C. schistoacanthus* presented in Table II illustrate this procedure. The density data (Peters, unpubl. data) were collected from 1.4 km of 10-m wide transects sampled at the Danau Sentarum Wildlife Reserve in the same area as the yield studies. At least three points of interest are illustrated by the data shown in Table II. The first is related to the extremely high density of the *C. schistoacanthus* populations at Danau Sentarum. Over 750 individual clumps per hectare were recorded in the transects, and, on the basis of stem counts made on a subsample of individuals, these clumps contained an estimated 12,200 canes/ha. The fact that *C. schistoacanthus* is one of the few Bornean rattans that can tolerate severe seasonal flooding is probably largely responsible for the notable abundance of this species.

The second point is that the total cane yield by the species is a function of both the size-specific growth rate and the density of canes in each size class. The larger size classes grow faster, but they represent a small percentage of the total annual productivity because of the limited number of clumps of this size in each hectare. Small clumps are very abundant, but they exhibit a growth rate that is less than a third of that shown by taller canes. Most

Table II. Estimated annual yield of rattan cane by a 1.0-ha population of *Calamus schistoacanthus* growing in seasonally flooded forest at the Danau Sentarum Wildlife Reserve, West Kalimantan, Indonesia

Height class (m)	Clumps/ha	Canes/ha[1]	Estimated mean growth (cm/yr)[2]	Total growth/class (m)
0.0–1.0	327	3177	25.1	797.4
1.0–2.0	174	2234	43.6	974.0
2.0–3.0	99	1679	62.3	1046.0
3.0–4.0	57	1277	80.9	1033.1
4.0–5.0	36	1066	80.9	862.3
5.0–6.0	32	1251	80.9	1012.1
6.0–7.0	18	930	80.9	752.4
7.0+	11	653	80.9	528.3
Total	754	12,267		7005.6

[1] Estimates of the number of canes (ramets) per clump (genet) were based on counts of a subsample of clumps ($n=46$) and calculated using the equation $\log_{10}$ (number of canes) $= a + b$ (height of cluster); $a=0.927$, $b=0.121$, $r^2=0.82$.
[2] Size-specific growth estimates were calculated using the regression equation growth $= a + b$ (height), where $a=15.76$, $b=0.186$, $r^2=0.78$.

of the rattan cane produced each year by the *C. schistoacanthus* population comes from individuals of intermediate size (2.0–6.0 m tall).

Finally, the data shown in Table II provide some indication about what a sustainable level of rattan harvest might be from these populations. An estimated total of 7005 m of cane per hectare are produced every year by *C. schistoacanthus*. Of this total, 3155 m of cane are produced by individuals of merchantable size (i.e. the four size classes >4.0 m tall). Given that the minimum length of harvested cane is usually 4.0 m, this figure represents a mean annual productivity of approximately 790 canes/ha. Every year, at least half of the individuals in the 3.0–4.0 m size class will grow into the 4.0–5.0 m merchantable class. The "ingrowth" of these smaller individuals expands the local rattan resource base by about 650–700 new merchantable canes. If we assume that the density estimates are representative and that the measured growth rates are maintained over time, the data in Table II suggest that about 700 *C. schistoacanthus* canes per hectare per year could be harvested on a sustained-yield basis from the flooded forest of Danau Sentarum.

Conclusions

Ethnobotany and plant ecology are natural partners, and their collaboration can contribute greatly to the study of people and plants. Coupling plant use information with quantitative data on the distribution, abundance, and yield of different resources provides a useful new framework for addressing the question, How important is this species? Perhaps of even greater relevance given current realities, however, is that this integrated focus also allows the investigator to probe deeper into the questions, How quickly is this resource being used up? What can be done to prevent overexploitation? The conservation and rational use of the innumerable plant resources "discovered" by ethnobotanists over the last 100 years will inevitably require the collection of density and yield data. Although the fact is seldom mentioned, ethnobotanical research is really the first step toward effective resource management. The more ambitious the first step, the faster effective resource management can be achieved.

Acknowledgments

I thank Miguel Alexiades for his patience and persistence in motivating me to write about the links between ecology and ethnobotany. Fieldwork in West Kalimantan, Indonesia, was conducted under the auspices of the Indonesian Institute of Science (LIPI) in collaboration with Tanjungpura University in Pontianak, and the continual support of those two institutes is gratefully acknowledged. Funding for the research in Indonesia was provided by the World Environment and Resources Program of the John D. and Catherine T. MacArthur Foundation. I thank Wim Giesen, Eddy Zulkarnain, Budi Suriansyah, and Julia Aglionby for their invaluable assistance during the rattan surveys at Danau Sentarum.

Literature Cited

Adlard, P. G. 1990. Procedures for monitoring tree growth and site change: A field manual. Tropical Forestry Paper No. 23. Oxford Foresty Institute, Oxford.

Alder, D. 1980. Forest volume estimation and yield prediction. FAO Forestry Paper 22/2. Food and Agriculture Organization of the United Nations, Rome.

Anderson, A. B., A. Gely, J. Strudwick, G. L. Sobel & M. G. C. Pinto. 1985. Um sistema agroforestal na várzea do estuário amazônica (Ilha das Onças, Município de Barcarena, Estado do Pará). Acta Amazonica Supplement **15:** 195–224.

Ash, J. 1987. Demography of *Cyathea hornei* (Cyathaceae), a tropical tree fern from Fiji. Australian Journal of Botany **35:** 331–342.

Avery, T. E. & H. E. Burkhart. 1983. Forest measurements. McGraw-Hill, New York.

Balée, W. L. 1994. Footprints of the forest: Ka'apor ethnobotany—The historical ecology of plant utilization by an Amazonian people. Columbia University Press, New York.

Barrera de Jorgenson, A. 1993. Chicle extraction and forest conservation in Quintana Roo, Mexico. Thesis. University of Florida, Gainesville.

Bonham, C. D. 1989. Measurements for terrestrial vegetation. John Wiley, New York.

Boom, B. M. 1989. Use of plant resources by the Chacobo. Advances in Economic Botany **7:** 78–96.

Bormann, F. H. 1953. The statistical efficiency of sample plot size and shape in forest ecology. Ecology **34:** 474–487.

———— **& T. T. Kozlowski.** 1962. Measurement of tree growth with dial gauge dendrometers and vernier tree-ring bands. Ecology **43:** 289–294.

Bullock, S. 1980. Demography of an undergrowth palm in littoral Cameroon. Biotropica **12**: 247–255.

Campbell, D. G., D. C. Daley, G. T. Prance & U. N. Maciel. 1986. Quantitative ecological inventory of terra firme and várzea tropical forest on the Río Xingu, Brazilian Amazon. Brittonia **38**: 369–393.

Chapman, C. A., L. J. Chapman, R. Wangham, K. Hunt, D. Gebo & L. Garder. 1992. Estimators of fruit abundance of tropical trees. Biotropica **24**: 527–531.

Clark, D. A. & D. B. Clark. 1987. Temporal and environmental patterns of reproduction in *Zamia skinneri,* a tropical rain forest cycad. Journal of Ecology **75**: 135–149.

Cochran, W. G. 1977. Sampling techniques. 2nd ed. John Wiley, New York.

Dinerstein, E. 1986. Reproductive ecology of fruit bats and the seasonality of fruit production in a Costa Rican cloud forest. Biotropica **18**: 37–318.

Dove, M. R. 1993. Smallholder rubber and swidden agriculture in Borneo: A sustainable adaptation to the ecology and economy of the tropical forest. Economic Botany **47**: 136–147.

Dransfield, J. & N. Manokaran (eds.). 1994. Rattans. Plant Resources of South-East Asia No. 6, Plant Resources of South East Asia Foundation, Bogor, Indonesia.

Food and Agriculture Organization (FAO). 1973. Manual of forest inventory with special reference to mixed tropical forest. Food and Agriculture Organization of the United Nations, Rome.

Fong, F. W. 1992. Perspectives for sustainable utilization and management of Nipa vegetation. Economic Botany **46**: 45–54.

Gauch, H. G. 1982. Multivariate analysis in community ecology. Cambridge University Press, Cambridge, U.K.

Gentry, A. H. 1982. Patterns of neotropical plant species diversity. Evolutionary Biology **15**: 1–84.

——. 1988. Tree species richness of upper Amazonian forests. Proceedings of the National Academy of Sciences **85**: 156–159.

—— (ed.). 1990. Four neotropical forests. Yale University Press, New Haven, Conn.

Gianno, R. 1986. The exploitation of resinous products in a lowland Malayan forest. Wallaceana **43**: 3–6.

Goldsmith, F. B. & C. M. Harrison. 1976. Description and analysis of vegetation. Pages 85–155 *in* S. B. Chapman, ed., Methods in plant ecology. John Wiley, New York.

Green, D. F. & E. A. Johnson. 1994. Estimating the mean annual seed production of trees. Ecology **75**: 642–647.

Greig-Smith, P. 1983. Quantitative plant ecology. 3rd ed. University of California Press, Los Angeles.

Hall, P. & K. Bawa. 1993. Methods to assess the impact of extraction of non-timber tropical forest products on plant populations. Economic Botany **47**: 234–247.

Harshberger, J. W. 1896. Purposes of ethnobotany. Botanical Gazette **21**: 146–154.

Heinsdijk, D. & M. A. de Bastos. 1965. Forest inventory on the Amazon. FAO Report No. 2080. Food and Agriculture Organization of the United Nations, Rome.

Hodge, W. H. 1948. Wartime *Cinchona* procurement in Latin America. Economic Botany **2:** 229–257.

Holdridge, L. R., W. C. Grenke, W. H. Hatheway, T. Liiaing & J. S. Tosi. 1971. Forest environments in tropical life zones: A pilot study. Pergamon Press, Oxford.

Howe, H. F. 1977. Bird activity and seed dispersal of a tropical wet forest tree. Ecology **58:** 539–550.

———. 1980. Monkey dispersal and waste of a neotropical tree. Ecology **61:** 944–959.

——— & G. A. Vande Kerckhove. 1981. Removal of wild nutmeg *(Virola surinamensis)* crop by birds. Ecology **62:** 1093–1106.

Hubbell, S. P. 1979. Tree dispersion, abundance and diversity in tropical dry forest. Science **203:** 1299–1309.

Husch, B., C. I. Miller & T. W. Beers. 1972. Forest mensuration. Ronald Press, New York.

Irvine, D. 1989. Succession management and resource distribution in an Amazonian rain forest. Advances in Economic Botany **7:** 223–237.

Kershaw, K. A. & J. H. H. Looney. 1985. Quantitative and dynamic plant ecology. 3rd ed. Edward Arnold, London.

Kinnard, M. F. 1992. Competition for a forest palm: Use of *Phoenix reclinata* J. J. Jacquin by human and non-human primates. Conservation Biology **6:** 101–107.

Knight, D. H. 1975. A phytosociological analysis of species-rich tropical forest on Barro Colorado Island, Panama. Ecological Monographs **45:** 258–254.

Köhl, M. 1993. Forest inventory. Pages 243–332 *in* L. Pancel, ed., Tropical forestry handbook. Springer-Verlag, Berlin.

Kozlowski, T. T., P. J. Kramer & S. G. Pallardy. 1991. The physiological ecology of woody plants. Academic Press, New York.

Lang, G. E., D. H. Knight & D. A. Anderson. 1971. Sampling the density of tree species with quadrats in a species-rich tropical forest. Forest Science **17:** 395–400.

Lepofsky, D. 1992. Arboriculture in the Mussau Islands, Bismark Archipelago. Economic Botany **46:** 192–211.

Liming, F. C. 1957. Homemade dendrometers. Journal of Forestry **55:** 575–577.

Lugo, A. & C. Rivera. 1987. Leaf production, growth rate, and age of the palm *Prestoea montana* in the Luquillo experimental forest, Puerto Rico. Journal of Tropical Ecology **3:** 1151–1161.

Lyon, L. J. 1968. An evaluation of density sampling methods in a shrub community. Journal of Range Management **21:** 16–20.

Mueller-Dombois, D. & H. Ellenberg. 1974. Aims and methods of vegetation ecology. John Wiley, New York.

Myers, E. & V. J. Chapman. 1953. Statistical analyses applied to a vegetation type in New Zealand. Ecology **34:** 175–185.

Oyama, K. 1990. Variation in growth and reproduction in the neotropical dioecious palm *Chamaedorea tepejilote*. Journal of Ecology **78**: 648–663.

Padoch, C. & W. de Jong. 1991. The house gardens of Santa Rosa: Diversity and variability in an Amazonian agricultural system. Economic Botany **45**: 1166–1175.

———— **& C. Peters.** 1993. Managed forest gardens in West Kalimantan, Indonesia. Pages 167–176 in C. S. Potter, J. I. Cohen & D. Janczewski, eds., Perspectives on biodiversity: Case studies of genetic resource conservation and development. AAAS Press, Washington, D.C.

Peters, C. M. 1990. Plant demography and the management of tropical forest resources: A case study of *Brosimum alicastrum* in Mexico. Pages 265–272 *in* A. Gomez-Pompa, T. C. Whitmore & M. Hadley, eds., Rain forest regeneration and management. Cambridge University Press, Cambridge.

————. 1994. Sustainable harvest of non-timber plant resources in tropical moist forest: An ecological primer. Biodiversity Support Program, Washington, D.C.

———— **& E. J. Hammond.** 1990. Fruits from the flooded forests of Peruvian Amazonia: Yield estimates for natural populations of three promising species. Advances in Economic Botany **8**: 159–176.

———— **& A. Vazquez.** 1987. Estudios ecologicos de camu-camu *(Myrciaria dubia)* I. Produccion de frutos en poblaciones naturales. Acta Amazonica **16/17**: 161–174.

Philip, M. S. 1994. Measuring trees and forests. 2nd ed. CAB International, Wallingford, U.K.

Phillips, O. 1993. The potential for harvesting fruits in tropical rainforests: New data from Amazonian Peru. Biodiversity and Conservation **2**: 18–38.

Pinard, M. 1993. Impact of stem harvesting on populations of *Iriartea deltoidea* (Palmae) in an extractive reserve in Acre, Brazil. Biotorpica **25**: 2–14.

Pinedo-Vasquez, M., D. Zarin, P. Jipp & J. Chota-Iuma. 1990. Use-values of tree species in a communal forest reserve in northeast Peru. Conservation Biology **4**: 405–416.

Piñero, D. & J. Sarukhan. 1982. Reproductive behavior and its individual variability in a tropical palm, *Astrocaryum mexicanum*. Journal of Ecology **72**: 977–991.

————, **J. Sarukhan & E. Gonzales.** 1977. Estudios demográficos en plantas. *Astrocaryum mexicanum* Lieb. I. Estructura de la poblaciones. Boletin de la Sociedad Botánica de Mexico **37**: 69–118.

Prance, G. T., W. Balée, B. M. Boom & R. L. Carneiro. 1987. Quantitative ethnobotany and the case for conservation in Amazonia. Conservation Biology **1**: 296–310.

Reid, N., J. Marroquin & P. Beyer-Munzel. 1990. Utilization of shrubs and trees for browse, fuelwood and timber in the Tamaulipan thornscrub, northeastern Mexico. Forest Ecology and Management **36**: 61–79.

Rico-Gray, V., J. G. Garcia-Franco, A. Chemas, A. Puch & P. Sima. 1990. Species composition, similarity, and structure of Maya homegardens in Tixpeual and Tixcacaltuyub, Yucatan, Mexico. Economic Botany **44**: 470–487.

Salick, J. 1989. Ecological basis of Amuesha agriculture, Peruvian upper Amazon. Advances in Economic Botany **7:** 189–212.

Sarukhan, J. 1978. Studies on the demography of tropical trees. Pages 163–184 *in* P. B. Tomlinson & M. H. Zimmerman, eds., Tropical trees as living systems. Cambridge University Press, Cambridge.

Snedecor, W. G. & G. W. Cochran. 1967. Statistical methods. 6th ed. Iowa State University Press, Ames.

Sokal, R. R. & F. J. Rohlf. 1981. Biometry. 2nd ed. W. H. Freeman, San Francisco.

Sork, V. L. 1987. Effects of predation and light on seedling establishment in *Gustavia superba.* Ecology **68:** 1341–1350.

Tanner, E. V. J. 1983. Leaf demography and growth of the tree fern *Cyathea pubescens* Mett. *ex* Kuhn in Jamaica. Journal of the Linnean Society, Botany **87:** 213–227.

UNESCO. 1978. Tropical forest ecosystems: A state-of-knowledge report. UNESCO, Paris.

Van Dyne, G. M., W. G. Vogel & H. G. Fisser. 1963. Influence of small plot size and shape on range herbage production estimates. Ecology **44:** 746–759.

Wan Razali, M., H. T. Chan & S. Appanah (eds.). 1989. Growth and yield in tropical mixed/moist forests. Forest Research Institute of Malaysia, Kepong.

Whittaker, R. H. 1967. Gradient analysis of vegetation. Biological Reviews **42:** 2077–2264.

Wood, G. B. 1989. Ground sampling methods used to inventory tropical mixed/moist forest: The challenge before us. Pages 51–59 *in* M. Wan Razali, H. T. Chan & S. Appanah, eds., Growth and yield in tropical mixed/moist forests. Forest Research Institute of Malaysia, Kepong.

IV
Appendices

Introduction

The preceding discussions have introduced a number of broad and complex topics relating to the theory, practice, and ethics of ethnobotanical fieldwork. We hope that they will help broaden the reader's perspective of ethnobotany and also provide a blueprint for developing new complementary approaches and skills to fieldwork. The following appendices seek to further assist this goal by providing a sampling of key resources dealing with botany, anthropology, and ethnobotany. Clearly, these bibliographical lists are not complete or all-inclusive, but they aim to illustrate a range of approaches and provide an effective platform from which to access additional information.

Appendix 1: Botany-Related Resources for Ethnobotanists

Compiled by M. N. Alexiades

Floras, florulas, checklists, monographs, and keys are important resources for ethnobotanists wishing to identify plants in the field. A review of published floristic treatments for different tropical regions is included in Campbell and Hammond, 1989. Unfortunately, most country floras are incomplete or unfinished. Widely used floras in the Neotropics include *Flora of Ecuador* (Harling & Sparre, 1973–present), *Flora of Guatemala* (Steyermark & Williams, 1946–1977), *Flora Costaricensis* (Burger, 1971–present), *Flora of Panama* (Woodson & Schery, 1943–1980), and *Flora of Peru* (Macbride, 1936–1961). Though largely incomplete, *Flora of Colombia* (Instituto de Ciencias Naturales, 1983–present) and *Flora of Venezuela* (Lasser, 1964–present) include many important families. In addition, fieldworkers should check for any regional floras, florulas, or checklists available for the region surrounding their study area.

Flora Neotropica, published by the Organization for Flora Neotropica of The New York Botanical Garden, contains monographic treatments of a number of neotropical plant families. There are also keys and field manuals to assist in the identification of plants, usually to family or genus only. Published keys and guides to neotropical plant families include Gentry, 1993; Maas et al., 1993; and Simpson and Janos, 1974. Heywood, 1993, provides a general account of the world's plant families. A detailed description of all flowering plant families and a discussion of their systematic and taxonomic relationships can be found in Cronquist, 1981.

Herbaria are another important resource for ethnobotanists. Herbarium specimens are valuable aids for the identification of plant specimens, and labels often include local names and uses (Von Reis, 1973). Furthermore, taxonomists working in herbaria may be interested in certain plant groups or have valuable advice or suggestions for study in areas with which they are familiar. *Index Herbariorum* (Holmgren et al., 1992) provides a systematic listing of all herbaria in the world, including their staff and specialties and important collections.

General introductions to plant biology and plant systematics can be found in Cronquist, 1988; Jones & Luchsinger, 1979; and Raven et al., 1992.

Literature Cited

Burger, W. C. (ed.). 1971–present. Flora Costaricensis. Fieldiana Botany **35.**

Campbell, D. G. & H. D. Hammond (eds). 1989. Floristic inventory of tropical countries: The current status of plant systematics, collections, and vegetation, plus recommendations for the future. The New York Botanical Garden, Bronx.

Cronquist, A. C. 1981. An integrated system of classification of flowering plants. Columbia University Press, New York.

———. 1988. The evolution and classification of flowering plants. 2nd. ed. The New York Botanical Garden, Bronx.

Gentry, A. H. 1993. A field guide to the families and genera of woody plants of northwest South America (Colombia, Ecuador, Peru) with supplementary notes on herbaceous taxa. Conservation International, Washington, D.C.

Harling, G. & B. Sparre (eds.). 1973–. Flora of Ecuador. University of Gotenburg, Lund, Sweden.

Heywood, V. H. (ed.). 1993. Flowering plants of the world. Rev. ed. Oxford University Press, New York.

Holmgren, P. K., N. H. Holmgren & L. C. Barnett (eds.). 1992. Index herbariorum. Part I: The herbaria of the world. 8th ed. The New York Botanical Garden, Bronx.

Instituto de Ciencias Naturales. 1983–present. Flora de Colombia. Universidad Nacional de Colombia, Bogotá.

Jones, S. B. & A. E. Luchsinger. 1979. Plant systematics. 2nd ed. McGraw-Hill, New York.

Lasser, T. 1964–present. Flora de Venezuela. Instituto Botánico, Caracas.

Maas, P. J. M., L. Y. Th. Westra & A. Farjon. 1993. Neotropical plant families: A concise guide to families of vascular plants in the Neotropics. Koeltz Scientific Books, Champaign, Ill.

Macbride, J. F. 1936–1961. Flora of Peru. Publications of the Field Museum of Natural History. Botanical Series 13.

Raven, P. H., R. F. Evert & S. E. Eichhorn. 1992. Biology of plants. 5th ed. Worth Publishers, New York.

Simpson, D. R. & D. Janos. 1974. Punch card key to the families of dicotyledons of the Western Hemisphere south of the United States. Field Museum of Natural History, Chicago.

Steyermarck, J. & L. O. Williams. 1946–1977. Flora of Guatemala. Fieldiana Botany **24.** 12 volumes.

Von Reis, S. 1973. Drugs and foods from little-known plants: Notes in Harvard University Herbaria. Harvard University Press, Cambridge, Mass.

Woodson, R. E. & R. W. Schery. 1943–1980. Flora of Panama. Annals of the Missouri Botanical Garden **30(2)–67(4).**

Appendix 2: Anthropology-Related Resources for Ethnobotanists

Compiled by M. N. Alexiades

Written Resources

Fieldworkers commencing research in an area should review the ethnographic literature pertinent to the groups with which they will be working. Linguistical studies and lexicons, when available, are important aids in facilitating correct transcription of plant names. Social Science Abstracts, library periodicals and book catalogs, and on-line literature databases all can help locate relevant literature. Westerman (1994) offers a practical guide on how to access bibliographical materials pertaining to anthropology, including databases, in the library and through electronic mail. The Human Relations Area Files (HRAF) database is the most important archive of ethnographic materials, providing information about more than 6000 books and articles covering over 340 cultural groups around the world. Most university and research libraries will have access to this database, a description of which is included in Bernard (1988). Historical accounts and chronicles can also provide information on local history and migrations relevant to an ethnobotanical study. For example, missionaries have been present in some areas for considerable periods, and their writings can often provide valuable historical and ethnographic information.

Training in Cross-Cultural Studies

A general introduction to anthropology is provided by Haviland (1990). Introductory texts to cultural anthropology include Harris, 1993, and Murphy, 1989. Basic courses in ethnography and in cultural, ecological, and medical anthropology may provide a useful background in the cultural aspects of ethnobotany. Good introductory treatments to medical anthropology and discussions of ethnomedicine—the study of human beliefs and practices concerning illness and the body—include Landy, 1977, 1983; Johnson & Sargent, 1990; McElroy & Townsend, 1989; Worsley, 1982; and Young, 1982. Introductory and general texts in ecological anthropology and human ecology include Hardesty, 1977, and Sponsel, 1986.

Ethnoscience, a branch of cultural anthropology, studies the use people rules to categorize their experience of the surrounding environment. Basic texts include Tyler, 1969; Werner & Schoepfle, 1987a,b; and Weller & Romney, 1988. Spradley (1979) provides an excellent review of interviewing techniques from an ethnoscientific perspective.

Training in Linguistic Skills

Good introductory texts to linguistics include Akmajian et al., 1990, and O'Grady et al., 1989. Articulatory phonetics deals with human sounds resulting from particular configurations of the articulatory apparatus. Most university linguistics departments offer courses. The Summer Institute of Linguistics (SIL; 7500 W. Camp Wisdom Road, Dallas, Texas 75236, USA) offers intensive 3-month courses in linguistics in several universities in the United States and Canada. Some textbooks are available (e.g., Smalley, 1964), but they are of limited use because developing the correct skills requires listening to the different sounds. Phonemics, on the other hand, is related to the technique used to reduce unwritten languages to written form. In addition to courses offered in linguistic departments, there are some manuals that can be used in the field (e.g., Pike, 1971). Orthographies (a system of writing or phonemization) are available for some languages. Missionary organizations such as the Summer Institute of Linguistics (SIL) have published a number of basic linguistic aids for several oral languages in the Neotropics. Missionaries or anthropologists who have conducted linguistic studies in the area in question may be a further source of technical assistance in transcribing indigenous plant names.

Literature Cited

Akmajian, A., R. A. Demers, A. K. Farmer & R. M. Harnish. 1990. Linguistics: An introduction to language and communications. 3rd ed. MIT Press, Cambridge, Mass.

Bernard, H. R. 1988. Research methods in cultural anthropology. Sage, Newbury Park, Calif.

Hardesty, D. L. 1977. Ecological anthropology. Wiley, New York.

Harris, M. 1993. Culture, people, nature: An introduction to general anthropology. Ed. 6. Harper Collins, New York.

Haviland, W. A. 1990. Anthropology. 6th ed. Holt, Rinehart and Winston, Chicago.

Johnson, T. M. & C. F. Sargent (eds.). 1990. Medical anthropology: A handbook of theory and method. Greenwood Press, Westport, Conn.

Landy, D. (ed.). 1977. Culture, disease, and healing: Studies in medical anthropology. Macmillan, New York.

————. 1983. Medical anthropology: A critical appraisal. Advances in Medical Science 1: 184–314.

McElroy, A. & P. K. Townsend. 1989. Medical anthropology in ecological perspective. Westview Press, Boulder, Colo.

Murphy, R. F. 1989. Cultural and social anthropology: An overture. 3rd ed. Prentice Hall, Englewood Cliffs, N.J.

O'Grady, W., M. Dobrovolsky & M. Aronoff. 1989. Contemporary linguistics: An introduction. St. Martin's Press, New York.

Pike, K. L. 1971. Phonemics: A technique of reducing language to writing. University of Michigan Press, Ann Arbor.

Smalley, W. A. 1964. Manual of articulatory phonetics. Practical Anthropology, Tarrytown, N.Y.

Sponsel, L. 1986. Amazon ecology and adaptation. Annual Review of Anthropology 15: 67–97.

Spradley, J. P. 1979. The ethnographic interview. Holt, Rinehart and Winston, New York.

Tyler, S. A. 1969. Cognitive anthropology. Holt, Rinehart and Winston, New York.

Weller, S. C. & A. K. Romney. 1988. Systematic data collection. Sage, Newbury Park, Calif.

Werner O. & G. M. Schoepfle. 1987a. Systematic fieldwork. Ethnographic analysis and data management. Vol. 1: Foundations of ethnography and interviewing. Sage, Newbury Park, Calif.

————— & —————. 1987b. Systematic fieldwork. Ethnographic analysis and data management. Vol. 2: Ethnographic analysis and data management. Sage, Newbury Park, Calif.

Westerman, R. C. 1994. Fieldwork in the library: A guide to anthropology and related area studies. American Library Association, Chicago.

Worsley, P. 1982. Non-western medical systems. Annual Review of Anthropology **11:** 315–348.

Young, A. 1982. The anthropologies of illness and sickness. Annual Review of Anthropology **11:** 257–285.

Appendix 3: Additional Ethnobotanical References

Compiled by M. Alexiades and J. Wood Sheldon
with suggestions from
W. Balée, M. J. Balick, B. Bennett, D. Daly,
E. Forero, C. Padoch, O. Phillips, and D. Williams

General/Theoretical

Berlin, B. 1992. Ethnobiological classification: Principles of categorization of plants and animals in traditional societies. Princeton University Press, Princeton, N.J.

Berlin, B., D. E. Breedlove & P. H. Raven. 1974. Principles of Tzeltal plant classification: An introduction to the botanical ethnography of a Mayan-speaking people of highland Chiapas. Academic Press, New York.

Ellen, R. F. 1986. Ethnobiology, cognition, and the structure of prehension: Some general theoretical notes. Journal of Ethnobiology **6**: 83–98.

Etkin, N. L. (ed.). 1986. Plants in indigenous medicine and diet. Redgrave, Bedford Hills, N.Y.

Ford, R. I. (ed.). 1978. The nature and status of ethnobotany. Anthropological Papers No. 67. Museum of Anthropology, University of Michigan, Ann Arbor.

Hays, T. E. 1974. Mauna: Explorations in Ndumba ethnobotany. Dissertation. University of Washington.

Hunn, E. 1982. The utilitarian factor in folk biological classification. American Anthropologist **84:** 830–847.

Johns, T. 1990. With bitter herbs they shall eat it: Chemical ecology and the human origins of diet and medicine. University of Arizona Press, Tucson.

Lewis, W. H. & M. P. F. Elvin-Lewis. 1977. Medical botany: Plants affecting man's health. John Wiley & Sons, New York.

Martin, G. 1994. Ethnobotany: A methods manual. Chapman and Hall, New York.

Schultes, R. E. & S. von Reis. 1995. Ethnobotany. Evolution of a discipline. Dioscorides Press, Portland, Oregon.

Compilations and Symposia

Balick, M. J. & H. J. Beck. 1990. Useful palms of the world: A synoptic bibliography. Columbia University Press, New York.

Bárcenas, A., A. Barrera, J. Caballero & L. Durán (eds.). 1982. Memorias Simposio de Etnobotánica, México D. F., 1978. Instituto Nacional de Antropología e Historia, México.

Ciba Foundation. 1990. Bioactive compounds from plants. Ciba Foundation Symposium 154, 20–22 February 1990, Bangkok. John Wiley, New York.

Jain, S. K., V. Mudgal, D. K. Banerjee, A. Guha, D. C. Pal & D. Das. 1984. Bibliography of ethnobotany. Botanical Survey of India. Dept. of the Environment, Howrah.

Posey, D. A. & W. L. Overal (eds.). 1990. Ethnobiology: Implications and applications. Proceedings of the First International Congress of Ethnobiology, Belém, Pará, July 1988. Museu Paraense Emílio Goeldi, Belém.

Prance, G. T. & J. A. Kallunki (eds.). 1984. Ethnobotany in the Neotropics: Proceedings. Advances in Economic Botany **1.** New York Botanical Garden, Bronx.

Rios, M. & H. Borgtoft Pedersen. 1991. Las plantas y el hombre: Memorias del Primer Simposio Ecuatoriano de Etnobotánica y Botánica Económica. Ediciones ABYA-YALA, Quito.

Toledo, V. M. (ed.). 1987. Simposio de etnobotánica. Perspectivas en Latinoamérica. Anales del IV Congreso Latinoamericano de Botánica, Vol. IV. Serie Memorias de Eventos Científicos Colombianos No. 46. Instituto Colombiano para el Fomento de la Educación Superior, Bogotá.

Ethnobotany and Medicinal Plants: Latin America and the Caribbean

Acero Duarte, L. E. 1979. Principales plantas utiles de la Amazonia Colombiana. Instituto Geográfico "Agustín Codazzi," Bogotá.

————. 1982. Ethnobotany in Latin America: Propiedades, usos y nominacion de especies vegetales de la Amazonia Colombiana. Corporacion Aracuara, Bogotá.

Alcorn, J. B. 1984. Huastec Mayan ethnobotany. Texas University Press, Austin.

Arenas, P. 1981. Etnobotánica Lengua-Maskoy. República de Argentina, Buenos Aires.

Ayensu, E. S. 1981. Medicinal plants of the West Indies. Reference Publications, Algonac, Mich.

Balée, W. L. 1994. Footprints of the forest: Ka'apor ethnobotany—The historical ecology of plant utilization by an Amazonian people. Columbia University Press, New York.

Bastien, J. W. 1987. Healers of the Andes: Kallawaya herbalists and their medicinal plants. University of Utah Press, Salt Lake City.

Boom, B. 1987. Ethnobotany of the Chacobo Indians. Advances in Economic Botany **4**.

Davis, E. W. & J. A. Yost. 1983. The Ethnobotany of the Waraoni of eastern Ecuador. Botanical Museum Leaflets **29**: 159–217.

Denevan, W. M. & C. Padoch, (eds.). 1987. Swidden-fallow agroforestry in the Peruvian Amazon. Advances in Economic Botany **5**: 8–46.

Duke, J. A. & R. Vasquez. 1994. Amazonian ethnobotanical dictionary. CRC Press, Boca Raton, Fla.

Forero P., L. E. 1980. Etnobotánica de las comunidades indígenas Cuna y Waunana, Chocó (Colombia). Cespedesia **9(33–34)**: 118–292.

García-Barriga, H. 1974. Flora medicinal de Colombia. Instituto de Ciencias Naturales, Universiadad Nacional de Bogotá, Bogotá.

Glenobski, L. L. 1983. The ethnobotany of the Takuna Indians Amazonas, Colombia. Universidad Nacional de Colombia, Bogotá.

Hernandez X., E. 1970. Exploración etnobotánica y su metodología. Colegio de Postgraduados, Escuela Nacional de Agricultura, SAG, Chapingo, México.

La Rotta, C. 1989. Especies utilizadas por la comunidad Miraña: Estudio etnobotánico. World Wildlife Fund, Bogotá.

Mendieta, R. M. & R. S. del Amo. 1981. Plantas medicinales del estado de Yucatán. Instituto Nacional de Investigaciones Sobre Recursos Bióticos, Xalapa.

Patiño, V. M. 1989. Bibliografía etnobotánica parcial comentada de Colombia y los Paises Vecinos. Instituto Colombiano de Cultura Hispanica, Bogotá.

Posey, D. A. & W. Balée. 1989. Resource management in Amazonia. Advances in Economic Botany **7**. The New York Botanical Garden, Bronx.

Schultes, R. E. & R. F. Raffauf. 1990. The healing forest: Medicinal and toxic plants of the northwest Amazonia. Dioscorides Press, Portland, Ore.

Toledo, V. M. 1982. La etnobotanica hoy. Revisión del conocimiento, lucha indígena, y proyecto nacional. Biótica **7(2)**: 141–150.

Ethnobotany and Medicinal Plants: North America

Beck, B. & S. S. Strike. 1994. Ethnobotany of the California Indians. Koeltz Scientific Books, Champaign, Ill.

Elmore, F. H. 1944. Ethnobotany of the Navajo. New Mexico University Press, Albuquerque.

Felger, R. S. & M. B. Moser. 1985. People of the desert and sea: Ethnobotany of the Seri Indians. University of Arizona Press, Tucson.

Ford, R. I. (ed.). 1986. An ethnobiology source book: The uses of plants and animals by American Indians. Garland, New York.

Nabhan, G. P. 1985. Gathering the desert. University of Arizona Press, Tucson.

Turner, N. J. 1995. Ethnobotany today in northwestern North America. Pages 264–289 *in* R. E. Schultes & S. von Reis, eds. Ethnobotany. Evolution of a discipline. Dioscorides Press, Portland, Oregon.

Ethnobotany and Medicinal Plants: Asia and the Pacific

Anderson, E. F. 1993. Plants and people of the Golden Triangle: Ethnobotany of the hill tribes of northern Thailand. Dioscorides Press, Portland, Ore.

Barrau, J. (ed.). 1963. Plants and the migrations of Pacific peoples: A symposium. Bishop Museum Press, Honolulu.

Bedi, S. J. 1978. Ethnobotany of the Ratan Mahal Hills, Gujarat, India. Economic Botany **32:** 278–284.

Conklin, H. C. 1954. The relation of Huanunoo culture to the plant world. Dissertation. Yale University, New Haven, Conn.

Cox, P. A. & S. A. Banack (eds.). 1991. Islands, plants and Polynesians: An introduction to Polynesian ethnobotany. Dioscorides Press, Portland, Ore.

Lassak, E. V. & T. McCarthy. 1983. Australian medicinal plants. Methuen, North Ryde, Australia.

Martin, M. A. 1971. Introduction a l'ethnobotanique du Cambodge. Editions du Centre national de la recherche scientifique, Paris.

Wightman, G. M., D. M. Jackson & L. V. V. Williams. 1991. Alawa ethnobotany: Aboriginal plant use from Minyerri, Northern Australia. Northern Territory Botanical Bulletin, 0314-1810, No. 11. Conservation Commission of the Northern Territory, Palmerston.

Ethnobotany and Medicinal Plants: Africa

Ayensu, E. S. 1978. Medicinal plants of West Africa. Reference Publications, Algonac, Mich.

Boulos, L. 1983. Medicinal plants of North Africa. Reference Publications, Algonac, Mich.

Heine, B. & I. Heine. 1988. Plant concepts and plant use: An ethnobotanical survey of the semi-arid and arid lands of East Africa. Part 1. Plants of the Chamus (Kenya). Verlag Breitenbach, Saarbrucken, Germany.

Iwu, Maurice M. 1993. Handbook of African medicinal plants. CRC Press, Boca Raton, Fla.

Johns, T. & J. O. Kokwaro. 1991. Food plants of the Luo of Siaya District, Kenya. Economic Botany **45:** 103–113.

Kokwaro, J. O. 1995. Ethnobotany in Africa. Pages 216–225 *in* R. E. Schultes & S. von Reis, eds. Ethnobotany. Evolution of a discipline. Dioscorides Press, Portland, Oregon.

Morgan, W. T. W. 1981. Ethnobotany of the Turkana: Use of plants by a pastoral people and their livestock in Kenya. Economic Botany **35:** 96–130.

Oliver-Bever, B. 1986. Medicinal plants in tropical West Africa. Cambridge University Press, New York.

Journals

Colombia Amazonica. Corporacion Araracuara, Bogotá.

Economic Botany. Journal of the Society for Economic Botany. The New York Botanical Garden, New York.

Ethnobotany. Journal of the Society of Ethnobotanists. S. K. Jain, editor. DEEP-Publicaions A-3/27A, DDA Flats, Paschim Vihar, New Delhi 10063 India.

Human Ecology. Plenum, New York.

Journal of Ethnobiology. Journal of the Society of Ethnobiology. Center for Western Studies, Flagstaff, Ariz.

Journal of Ethnopharmacology. Elsevier Sequoia, Lausane, Switzerland.

Journal of Herbs, Spices and Medicinal Plants. Food Production Press, Binghamton, N.Y.

Index